Geography on the EDGE

Year 12 NCEA Level Two

Martin Newton and Justin Peat

NELSON
CENGAGE Learning·

Australia • Brazil • Japan • Korea • Mexico • Singapore • Spain • United Kingdom • United States

Geography on the Edge Level 2
1st Edition
Martin Newton
Justin Peat

Cover designer: Cheryl Smith, Macarn Design
Text designer: Cheryl Smith, Macarn Design
Production controller: Siew Han Ong

Any URLs contained in this publication were checked for currency during the production process. Note, however, that the publisher cannot vouch for the ongoing currency of URLs.

Acknowledgements

Illustrations on pages 20, 21 (bottom), 22 (bottom), 28, 36, 38, 72,93, 94, 95, 97, 98, 105, 237, 241, 242, 243 (middle right), 244, 255, 259, and 266 courtesy of Jamie Laurie.

Images on front cover, pages 4-5, 6, 10, 11, 13, 17, 24,25, 26, 27, 29, 33, 34, 39 (top left and right, middle right, bottom), 41, 45, 49, 51, 59, 63 (all), 56, 57, 66-67, 68, 71 (right), 73, 92, 120, 121, 122 (all), 140, 152, 153, 173, 154, 164, 172, 176 (top, middle), 184, 187, 188, 190, 191, 194, 195, 196, 198, 200, 202, 207, 215 (top right), 222, 223, 226 (right), 230-31, 233, 236, 240, 243 (top, bottom), 249, 250, 251, 252,253 (bottom), 262, 264, and 265 courtesy of Shutterstock.

Topographical maps on pages 7, 35, 213, 217 and 220 courtesy of LINZ. Images on pages 14, 18, 52 and 53 (middle) courtesy of GNS. Image on page 39 (middle left) courtesy of Minden Pictures. Image on page 54 courtesy of One Rings Tours. Images on pages 61, 62 and 63 courtesy of Glentanner Park (Mount Cook) Ltd. Images on 69 and 100 courtesy of GeoEye. Ayers Rock-Yulara Tourist Resort courtesy of Voyages. Images on pages 71 (left), 77, 98, 103, 108 (top), 116, 119 courtesy of Tourism Australian. Map on page 84 courtesy of Australian Government Bureau of Meteorology. Cartoon on page 150 courtesy of Creators.com. Death of Detroit infographic on page 151 courtesy of Raymmar Tirado. Image on page 174 courtesy of the New Zealand Herald.

For product information and technology assistance,
in Australia call **1300 790 853**;
in New Zealand call **0800 449 725**

For permission to use material from this text or product, please email
aust.permissions@cengage.com

National Library of New Zealand Cataloguing-in-Publication Data
Peat, Justin.
Geography on the edge : level 2 / Justin Peat and Martin Newton.
Includes bibliographical references and index.
ISBN 978-0-17-023331-6
1. Geography—New Zealand—Problems, exercises, etc. 2. Geography—Problems, exercises, etc. I. Newton, Martin, 1948- II. Title.
910.76—dc 23

Cengage Learning Australia
Level 7, 80 Dorcas Street
South Melbourne, Victoria Australia 3205

Cengage Learning New Zealand
Unit 4B Rosedale Office Park
331 Rosedale Road, Albany, North Shore 0632, NZ

For learning solutions, visit **cengage.com.au**

Printed in China by China Translation & Printing Services.
1 2 3 4 5 6 7 18 17 16 15 14

Contents

Large New Environment: Island Mountains

Zealand Natural Glaciated South and High Country

Glaciated South Island Mountains and High Country

A large New Zealand natural environment

Iconic feature of the Southern Alps and High Country: Aoraki/Mount Cook

Important understanding: Aoraki/Mount Cook is New Zealand's most famous and treasured landform feature. It is a feature located in the heart of the Southern Alps and High Country environment.

Figure 1 Aoraki/Mount Cook from the Hooker Valley.

 ISBN: 9780170233316

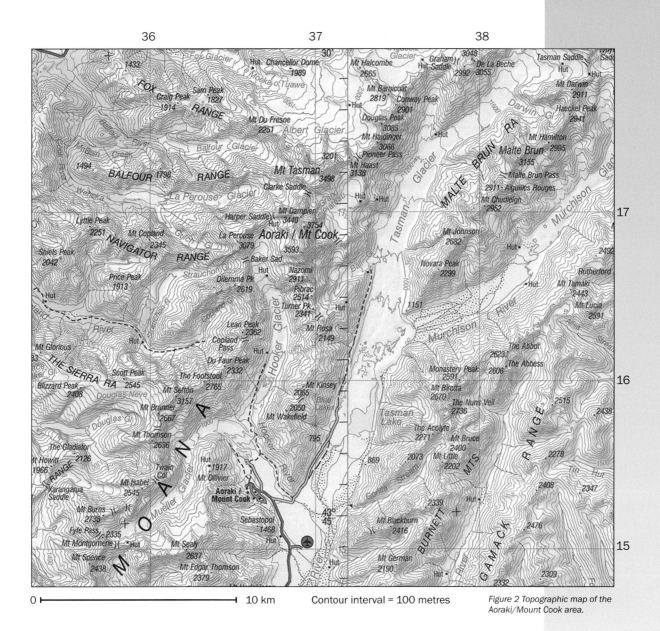

Figure 2 Topographic map of the Aoraki/Mount Cook area.

0 ⊢━━━━━━━━━━━━━━━⊣ 10 km   Contour interval = 100 metres

ROADS AND TRACKS [1]

State highway	①
Four lanes or more	
Two lanes (includes passing lanes)	
Narrow road	
Vehicle track	
Foot track	
Closed track (see warning note below) [2]	
Poled route	
Road surface — sealed	
Road surface — metalled	
Road surface — unmetalled	
Tunnel, tunnel under road	
Bridge; two lane, one lane	
Ford	
Gate, locked gate, cattlestop	
Footbridge, cableway or handwire [3]	

VEGETATION FEATURES

Native forest	
Exotic coniferous forest	
Exotic non-coniferous forest	
Scrub	
Scattered scrub	
Shelter belt	
Trees	
Orchard or vineyard	
Mangroves	

MISCELLANEOUS

Residential area	
Large buildings	
Isolated building	
Homestead, stockyard	· Awapuni ⊞
Glasshouse or greenhouse	
Church, cemetery, grave	
Training track	
Golf course, helipad	
Historic Māori pa, redoubt, monument, plaque or signpost	⩎ ⊠ ▲ ★
Reservoir covered, reservoir uncovered, tank	○ ○ ○
Mast, tower, wind machine or wind turbine	△ ⚡ ⵗ
Shipwreck, lighthouse, beacon	⊥ ⎕ ⍗
Fence (selection only)	
Pipeline above ground	
Pipeline underground	
Disused water race	
Power line on pylons (actual positions) [3]	
Power line on poles (away from roads) [3]	
Telephone line (away from roads) [3]	

RELIEF FEATURES [4]

Index contour	
Intermediate contours	
Perennial snow and ice contours	
Supplementary contour	
Depression contours	
Shallow depression, small depression or shaft	
Beaconed trig station (with trig identification code)	▲ A1B2
Elevation in metres	▲ 130m · 130
Cliff, terrace, slip	
Rock outcrops	
Stopbank, cutting	
Embankment or causeway	
Saddle, cave	⋉ ▲
Alpine features	
Moraine	
Moraine wall	
Scree	

Tribal links — Aoraki and Ngai Tahu

Aoraki is a special mountain for Ngai Tahu, who are the main South Island **iwi**. They tell the story of how Aoraki, the young son of Rakinui, the sky father, along with his brothers, brought his great **waka** down from the heavens in order to visit their stepmother, Papatuanuku (Mother Earth). They explored the earth and the seas, sailing their powerful waka through the dark empty southern oceans. Shortly after this, Aoraki and his brothers became hungry and began fishing. They were unsuccessful and decided to return to the heavens. When attempting this return, Aoraki misquoted his **karakia** and the canoe fell back into the water and turned over onto its side. As the brothers moved onto the back of the overturned canoe, the southerly wind froze them and turned them to stone. They remain there today as the main peaks in the Southern Alps, with Aoraki being the highest. It is for this reason that Ngai Tahu often call the South Island 'Te Waka o Aoraki' and the Southern Alps '**Ka Tiritiri o te Moana**' (frothing waters of the ocean).

To Ngai Tahu, Aoraki represents the most sacred of ancestors, the ancestor from whom they descend and who provides the iwi with a sense of identity, solidarity and purpose. The ancestor **embodied** in the mountain remains the physical sign of Aoraki, the link between the **supernatural** and the natural world. Ngai Tahu believe their association with the mountain provides strength to their culture, and **mana** to their status as tangata whenua — the people of the land.

Aoraki is a **tapu**, or sacred, mountain for Ngai Tahu, and there are restrictions on the way it is used. Generally, Maori would not climb to the summit of tapu mountains. Ngai Tahu explain that standing on the very top peak of Aoraki would **denigrate** its tapu status. Aoraki is viewed with fear, **awe** and respect by Ngai Tahu because it is the place of the **atua** and other spirits. The bones of high-ranking men and women were laid to rest in burial caves on the mountain. Songs, poetry and speeches on the marae of Ngai Tahu are full of references to Aoraki.

Figure 3 Carving of Aoraki and his brothers in the Aoraki/Mount Cook visitor centre.

Learning Activities

1   Refer to Figures 1 and 2.

a   Draw a precise sketch of Figure 1 and label Aoraki/Mount Cook, sharp ridges (arêtes), U-shaped Hooker Valley, Hooker River, areas of snow and ice, vegetated/scrub areas, scree slopes.

b   How many peaks above 3000 metres in height are marked and named on the map?

c   Give the six-figure grid reference for Aoraki/Mount Cook mountain peak.

d   Name a cultural feature shown on the map, and give the six-figure grid reference of this feature.

e   Name the natural feature located at 373158.

f   How far is it from the northern tip of Tasman Lake to the peak of Aoraki/Mt Cook?

g   In which direction was the camera pointing when the photo in Figure 1 was taken?

h   Use the contour lines to estimate the height of the hut in the Hooker Valley at 367162. How much vertical climb is there between here and the summit of Aoraki/Mount Cook?

i   Imagine you were tramping up the track on the western side of the Hooker Valley towards Aoraki/Mount Cook. Describe the landscape you would pass through and see in front of you.

2   Refer to Figure 3 and the 'Tribal links — Aoraki and Ngai Tahu' information.

a   Write definitions of these words (bold in the text): iwi, waka, karakia, Ka Tiritiri o te Moana, embodied, supernatural, mana, tapu, denigrate, awe, atua.

b   What is it that makes Aoraki a special mountain for Ngai Tahu? Answer by using a star diagram, a labelled drawing or writing a paragraph.

c   Write three or four sentences describing what is shown in the Mount Cook visitor centre carving.

d   Explain the origin of Aoraki as the highest South Island mountain according to the Ngai Tahu story/legend.

Aoraki/Mount Cook fact file

- Aoraki/Mount Cook is the highest mountain in New Zealand. It is located in the Aoraki/Mount Cook National Park. There are 25 peaks over 3000 metres in height in the park. The park is named after its highest peak.
- 'Aoraki' is the original Ngai Tahu name for the mountain.
- The name 'Mount Cook' was given to the mountain in 1851 by a British Royal Navy sea captain who was surveying the New Zealand coastline. He named the mountain in honour of Captain James Cook, who sailed around New Zealand in 1770. Cook never saw the mountain that was later to be named after him.
- In 1998, as part of the settlement of a Waitangi claim between Ngai Tahu and the Crown, the original Ngai Tahu name for the mountain of Aoraki was recognised. The official name of the mountain became 'Aoraki/Mount Cook'.
- Although Aoraki/Mt Cook is the highest New Zealand mountain (3754 metres), in global terms it is not a giant — it is less than half the height of many Himalayan peaks which are over 8000 metres high (see pages 232–233).
- Aoraki/Mt Cook and other high parts of the Southern Alps are formed out of a metamorphic rock called schist. Aoraki/Mt Cook schist is very old. It began to be formed beneath the sea 200–250 million years ago when mud and sand were deposited on the sea floor. This mud and sand became compressed into a dark and hard sedimentary rock called greywacke. Over time, heat and pressure changed (metamorphosed) the greywacke into a new rock type called schist.

- Aoraki/Mount Cook has been pushed up from beneath the sea by tectonic pressure and land movements along the Alpine Fault. The rock forming the summit of Aoraki/Mount Cook was below sea level less than a million years ago. At the same time the mountain was being pushed up, snow, ice and weathering were wearing down and reshaping the mountain into its present shape. These processes continue today — uplift of 5–10 mm takes place each year, but this is balanced by erosion losses.

Figure 4 The Mount Cook lily.

- Aoraki/Mount Cook is constantly changing. On 14 December 1991, part of the east face of Aoraki/Mount Cook collapsed and tumbled down the east side of the mountain. This huge avalanche of rock, ice and snow ended up deposited across the Tasman Glacier in the valley below. In a few seconds the avalanche reduced the height of Aoraki/Mount Cook by 10 metres from 3764 to 3754 metres.
- Studies made of the collapsed east face of the mountain and of the avalanche debris revealed that the schist rock that makes up the mountain, although hard, is also shattered and crumbly. Geologists described the rock as being like Weet-Bix.
- The summit of Aoraki/Mount Cook has a permanent snow cover. Many glaciers originate in the area below the summit and then flow out through U-shaped valleys to lower land on the western and eastern sides of the mountain.
- Storms, high rainfall, low temperatures and gale-force winds are features of the climate of this area. Aoraki/Mount Cook has an annual rainfall of 8000 mm (8 metres). At Aoraki/Mount Cook village, annual falls of 4000 mm are recorded. Much of this 'rain' falls as snow.
- The summit and high slopes of Aoraki/Mount Cook are covered in snow, ice and rock; only the lower slopes and heads of the valleys that lead up to the mountain are vegetated. Low alpine plants are the main form of vegetation — there is no forest cover. The most famous of the alpine plants is the Mount Cook lily. It is not a lily at all but is the largest buttercup in the world (Figure 4).
- Kea, falcons/karearea and black-backed gulls/karoro fly above the slopes of the mountain, but only the small rock wren/piwauwau survives the winter in high rock basins.
- 25 December 1894 is the date of the first recorded successful climb of Aoraki/Mount Cook. The climb was completed by three men — Jack Clarke, Tom Fyfe and George Graham.
- The main climbing season on Aoraki/Mount Cook is summertime between November and April.
- Aoraki/Mount Cook Village is 16 km south of the mountain. The village is a tourist centre and base camp for climbers.

Mountain	Height (metres)	Country/region	
Punkak Jaya	5040	New Guinea	Highest mountain in Oceania
Kosciusko	2228	Australia	Highest in Australia
Aoraki/Mount Cook	3754	New Zealand (SI)	Highest in New Zealand
Tasman	3497	New Zealand (SI)	Second highest in New Zealand
Aspiring/Tititea	3033	New Zealand (SI)	Highest in Mount Aspiring National Park
Ruapehu	2797	New Zealand (NI)	Highest in North Island
Taranaki/Mount Egmont	2518	New Zealand (NI)	Second highest in North Island
Hikurangi (East Coast)	1754	New Zealand (NI)	Highest non-volcanic North Island mountain

Figure 5 High mountains in Oceania.

 ISBN: 9780170233316

Figure 6 Looking across Lake Pukaki to Aoraki/Mount Cook.

Learning Activities

1 Refer to Aoraki/Mount Cook fact file.

Only two of the 10 following statements about Aoraki/Mount Cook are correct. Copy out the 10 statements but make appropriate corrections to the eight with inaccuracies.

a 'Aoraki' is the Ngai Tahu name for the mountain.

b Captain James Cook named the mountain in honour of his voyage around New Zealand in 1770.

c Aoraki/Mount Cook is 3754 metres higher than New Zealand's next highest mountain.

d River and wave erosion are rapidly wearing down the mountain.

e Aoraki/Mount Cook is made up of hard sedimentary rock called greywacke.

f Volcanic activity has created the mountain over a period of thousands of years.

g Aoraki/Mount Cook is one of many peaks over 3000 metres high in the Aoraki/ Mount Cook National Park.

h In summer, snow melts off the mountain and alpine plants like the Mount Cook lily become visible around the summit.

i Aoraki/Mount Cook is the highest mountain in Oceania.

j The main climbing season on the mountain is the winter months between November and April.

2 Construct a diagram or graph to show the 'highest mountain' information given in Figure 5. A bar graph (using one bar for each mountain) would be one method to use.

3 Draw a labelled sketch of Figure 6. Include these labels on your sketch (some have been included on the photograph to help you): Aoraki/Mount Cook, Lake Pukaki (ribbon lake), Tasman Valley, sharp ridges (arêtes), high mountain peaks (horn peaks), steep slopes, river flats, low grasses and tussock land, loose rock and shingle (scree) slopes, road/track.

Lucky to be alive 11 April 2006

Mark Hateley, a 23-year-old Australian, had just completed a tough 10-hour climb to the summit of Aoraki/Mount Cook. After posing for a photograph on the summit, he began his descent. The climb down suddenly turned into a nightmare. Just 50 metres from the summit and only a few minutes into his return journey, he lost his footing and tumbled out of control down the mountainside. He is lucky to still be alive.

He fell more than 100 metres and escaped death by rolling over an ice bridge across a 3-metre-wide and 20-metre-deep crevasse. If he had fallen into the crevasse he would almost certainly have been killed. Rescuers found Mr Hateley semi-conscious half an hour later. He was helicoptered off the mountain and rushed to Christchurch Hospital with a dislocated ankle, fractured fibula, broken tailbone and fluid on the lungs. He spent a week in intensive care and three weeks in a general ward before returning to Brisbane.

Mount Cook rescue team leader Aaron Halstead said Mr Hateley was extremely fortunate to have survived the fall. 'This guy was very, very lucky, in the location he fell and in the fact that the crevasse just happened to have a ledge on it, which it normally never does. If he had slipped higher up or lower down, he would have kept going,' said Mr Halstead. He had been told that as Mr Hateley fell, he tried to jam his ice axe into the mountain to self-arrest, a normal climbing technique. 'He had one go at it and then, boom – you're going terminal velocity just about within a couple of seconds,' said Mr Halstead. He said if Mr Hateley had not been slowed by the temporary ice ledge, he may have gone over the Gun Barrels – cliffs that are 700 m high. Mr Halstead said Mr Hateley not only had the fortune of landing on the tiny ledge but also had perfect timing as the ledge he rolled over was almost never there. 'If he'd fallen off that same spot two months earlier, he would have just kept going.' In March 2004, another Australian climber, Phil Toms, fell in almost exactly the same spot. However, with no ledge to stop him, he fell 1000 m and was killed.

Mr Halstead said Mount Cook – which has claimed more than 200 lives in the past 90 years – could be dangerous. Around 100–150 people reach the summit each year and thousands more climb in the surrounding area. Most do so without incident. 'There isn't a lot of room for error – you can't afford to be lazy and switch off. You have to be focused all the time. Mountaineering is risky but it's a calculated risk – having judgement to weigh up both the conditions and your ability at the time is vital. In the wrong conditions, Cook can be a dangerous mountain but it's all relative.'

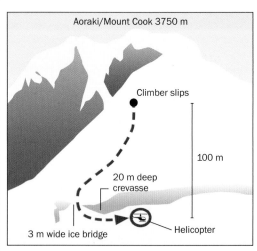

Figure 7 Scene of the disaster.

Mr Hateley only vaguely remembers the moments before his 10.30 a.m. fall, but friends have helped him piece together what happened. 'The ice had just shattered under my foot and once you lose your footing up in that situation it's obviously very slippery ... it was almost impossible to stop once you've started.' He has nothing but praise for his rescuers and the medical staff who treated him. 'Everyone who helped me out, I absolutely thank them,' he said. Despite his near-death experience, Mr Hateley is determined to climb again. 'Rock climbing I certainly will do. High mountains like that – we'll just have to see,' he said.

Learning Activities

1 How long did it take for Mark Hateley to climb to the top of Aoraki/Mount Cook?

2 Why is Mark Hateley lucky to be alive?

3 Retell the story of Mark Hateley's climb on Aoraki/Mt Cook in the first person. Write at least 100 words.

4 Draw a sketch of Figure 7 and add more annotations about the mountain and what happened to Mark Hateley during his climbing expedition on the mountain.

Crevasses

A crevasse is an open and deep vertical crack in a glacier or icefield. Crevasses can be several metres wide and up to 70 metres deep. Crevasses are the result of ice flow (movement) being quicker in some places than in others. This uneven movement causes tension, stretching and straining within the ice. Different flow speeds often occur at different depths within the ice and when a glacier moves around a bend or across an uneven and bumpy below ground surface. Crevasses are not permanent features — some are closing up while others are opening and this adds to their danger. Crevasse danger is increased after snowstorms, when bridges can develop on the surface and hide the crevasse below. The bridges are not always strong enough to support the weight of climbers, who may be unaware of the crevasse below.

A

B

Figure 8 a and b Climbing in crevassed *mountain terrain.*

Learning Activities

1 What is a crevasse?

2 Explain how crevasses form.

3 What is an ice bridge?

4 What makes crevasses so dangerous for climbers?

ISBN: 9780170233316

The Southern Alps and High Country environment: a mountainous and glaciated region

Important understanding: A natural environment is a distinctive part or area of the earth's surface with features that make it different from other areas. The Southern Alps and High Country is New Zealand's largest natural environment area. It is a high-altitude (mountainous) environment found in the South Island that has been affected by glaciation. The environment has many distinctive natural features (characteristics), including landforms, climate and vegetation. It is an environment treasured and valued by people for many different reasons, both in the past and present.

Across the Southern Alps and High Country

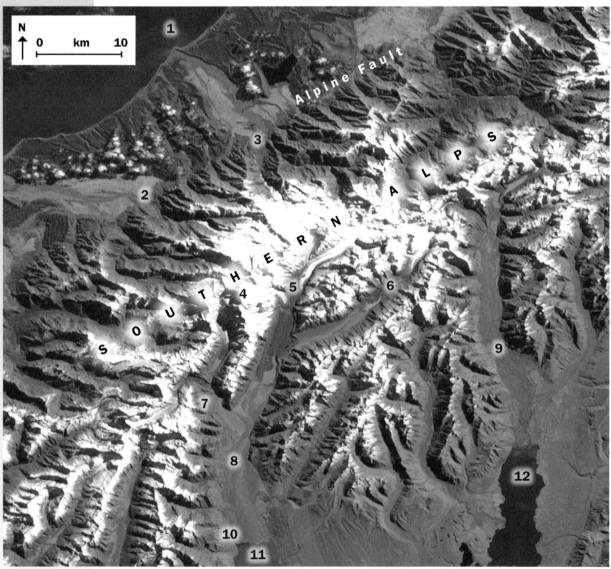

Figure 9
A satellite view across the Southern Alps and High Country.

1	Tasman Sea	**5**	Tasman Glacier	**10**	Glentanner Farm and Park Centre
2	Fox Glacier Village	**6**	Murchison Glacier	**11**	Lake Pukaki
3	Franz Josef Glacier Village	**7**	Mount Cook Village	**12**	Lake Tekapo
4	Aoraki/Mount Cook	**8**	Tasman River		
		9	Godley River		

 ISBN: 9780170233316

This large natural environment covers around one-third the area of the South Island. It is an area of rugged and elevated land that has been affected by glaciation. As well as being large in size, the environment has a range of other important and distinctive characteristics:

- spectacular and distinctive landforms; these are the result of internal tectonic forces, which have faulted, folded and lifted the land upwards, with external processes including glaciation then reshaping the land
- a climate that is cold, wet and windy for much of the year but also one that is changeable and shows huge contrasts in temperature and precipitation (snow and rainfall) within the region
- plants and animals that are found nowhere else in New Zealand
- being the source area of large South Island rivers that flow out from the mountains to the west and east coasts
- a farming region famous for large sheep stations but now an area of increasing farming diversity and intensification
- the destination area for huge numbers of domestic and overseas tourists; includes the Alps, Fiordland and the Queenstown region, which by most measures is the premier New Zealand tourism area
- producing and supplying the country with electricity from valley- and basin-located hydro-electric power stations.

Putting a boundary on the Southern Alps and High Country region

Figure 10 Satellite image of the South Island.

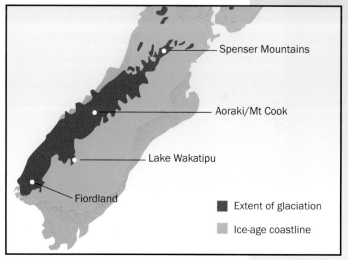

Figure 11 The last glaciation, 20,000–18,000 years ago.

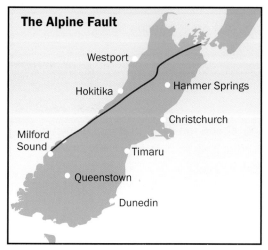

Figure 12 Alps, High Country and the Alpine Fault.

The location, size and extent of the Southern Alps and High Country environment is shown in Figure 11. It is the blue-shaded 'Extent of glaciation' area. This is an area of mostly high elevation (above 1000 metres) that was covered by ice during the last ice age.

This environment area extends from Fiordland in the southwest of the South Island, northeastwards along the main mountain divide as far as the Spenser Mountains between Kaikoura and Westport.

ISBN: 9780170233316

As well as the higher peaks, ridges and slopes of the Southern Alps, the environment also includes upper valley areas of rivers like the Clutha, Waitaki, Rakaia and Waimakariri and lakes including Wakatipu, Wanaka, Pukaki and Tekapo. Lower areas down to sea level along the West Coast are included within the boundaries of the environment because ice covered these areas during the last ice age and glaciers flowed down to sea level here.

Viewed from space, the Alpine Fault (Figure 12) is a distinctive feature of the landscape. This huge fault has had a major influence on landform development in the area.

Learning Activities

1   Refer to pages 14–16. Copy and complete the diagram below by adding two examples of natural features and two examples of cultural features that make the Southern Alps and High Country environment so distinctive.

2   Refer to the satellite image in Figure 9.

   a   The Southern Alps and the Alpine Fault have a similar orientation (direction and alignment). Is the orientation:

      i   N–S,

      ii   SE–NW

      iii SW–NE, or

      iv   W–E?

   b   In which direction does the Godley River flow?

   c   Name the lake the Godley River feeds into.

   d   Complete this sequence: Murchison Glacier ⟶ ? ⟶ Lake Pukaki.

   e   Name the settlement that is 15 km northwest of the peak of Aoraki/Mount Cook.

   f   How far is it from Mount Cook Village to the summit of Aoraki/Mount Cook?

   g   How far is it from the northwest corner of Lake Pukaki (by Glentanner Farm and Park Centre) to the Tasman Sea coast?

   h   What would make overland travel between these two places difficult?

   i   In this image, what direction are the slopes that are in shadow facing?

   j   Test your understanding:

      Explain why some slopes are in shadow when others have sunlight on them. A diagram could help your explanation.

3   Refer to Figures 10, 11 and 12.

   a   On a large outline map of the South Island, shade the blue 'Extent of glaciation' area from Figure 11. Label this shaded area 'Size and extent of the Southern Alps and High Country natural environment'.

   b   Add to the map the place names from Figure 11 and the Alpine Fault from Figure 12.

   ISBN: 9780170233316

Learning Activities

4 Copy and match the definitions in column B with the correct term in column A.

A Terms	B Definitions
Rugged	Raised-up land; high above sea level
Elevated	Broken earth crust, caused by internal pressures
Precipitation	A very large farm usually found in hill and mountain country
Glaciation	Rocky, rough and uneven land
Sheep station	Electricity produced in power stations that use water flow as their energy source; abbreviated as HEP
Hydro-electric power	A time of freezing cold when ice moves across and changes the land
Fault	Moisture from the sky, mainly rainfall and snowfall

5 Using no more than 40 words, describe the location, size and extent of the Southern Alps and High Country natural environment region.

Characteristics of the Southern Alps and High Country environment: natural elements with a human contribution

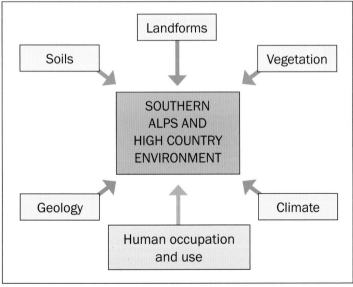

Figure 13 The elements that make up the environment.

Learning Activities

1 Make a copy of Figure 13 and include a key for the two different-coloured arrows used in the diagram.

2 How many natural elements make up the Southern Alps and High Country environment?

Figure 14 High mountain landforms.

Geology and landform elements of the environment — how the Southern Alps and High Country were formed and changed over time

Important understanding: The Southern Alps and High Country area is made up of different types of rock, with greywacke and schist the two most common ones.

A: Geology

Greywacke   Schist

Figure 15 Greywacke and schist rocks.

Rocks of the Southern Alps are ancient. Three main rock types — igneous, sedimentary and metamorphic — are present in the Southern Alps and High Country area. These rocks are hundreds of millions of years old but have been pushed up above sea level only during the last 5 million years.

Greywacke (Figure 15) is a dark-grey sandstone **sedimentary rock**. The sediments that make up the greywacke began their life as granite rock. The granite was weathered, eroded, transported by rivers and finally deposited as grains of quartz and feldspar on the sea floor 200 to 300 millions of years ago. Over time, these sediments became buried deeper and deeper, resulting in pressure that cemented them into a very hard rock. This rock is called greywacke.

The greywacke buried most deeply beneath the sea and covered by more and more sediments was put under greater pressure, and an increase in temperature, so that different minerals like shiny biotite grew within the rock and a new strongly layered rock, called **schist**, was formed. Schist is metamorphosed greywacke (Figure 15).

Most of the Southern Alps and High Country east of the Alpine Fault on the Pacific Plate is made of greywacke and schist (Figure 18).

Granite is an **igneous rock** and is found on the eastern edge of the Australian Plate to the west of the Alpine Fault (Figure 18). These rocks are older and harder than those of the Pacific Plate rocks. Some are 600 million years old.

Since these old rocks (granite, greywacke and schist) were formed, New Zealand has had periods where what is land today was buried beneath the sea, for example between 80 and 15 million years ago. During these times the old rocks were covered by sediments and '**new sedimentary rocks**' like sandstone, mudstone and limestone were formed. There are very few remains of these new rocks in the Southern Alps today because they have either been buried as the land has been compressed and shifted upwards and downwards or eroded away by the action of waves, ice and rivers.

 ISBN: 9780170233316

B: Land from beneath the sea — tectonic processes and the rise of the Southern Alps

Important understanding: Processes operating beneath the crust of the earth over millions of years have led to the formation of the Southern Alps and High Country.

Zealandia — New Zealand's continent

The crust of the earth is made up of two main types of rock. The ocean floor is made up of heavy basalt rock, while the continents are made of lighter granite rock. Because granite is lighter than basalt, it is more buoyant, so just like ice floats in water, the granite rocks float on top of the basalt (Figure 19). If you were to drill down through the continents (granite rock), there would be ocean crust (basalt rock) beneath.

Recently geologists have identified and mapped a huge area of previously unknown continental (granite) crustal rock in the Southwest Pacific. Geologists have named this area Zealandia and suggested it be recognised as a continent. This is New Zealand's continent, and is about half the size of Australia. Unlike the other continents, Zealandia is mostly beneath the sea. Zealandia extends from beyond Campbell Island to the south of New Zealand northwards to New Caledonia. New Zealand is located in the middle of Zealandia and is the largest part of Zealandia above sea level (Figure 16).

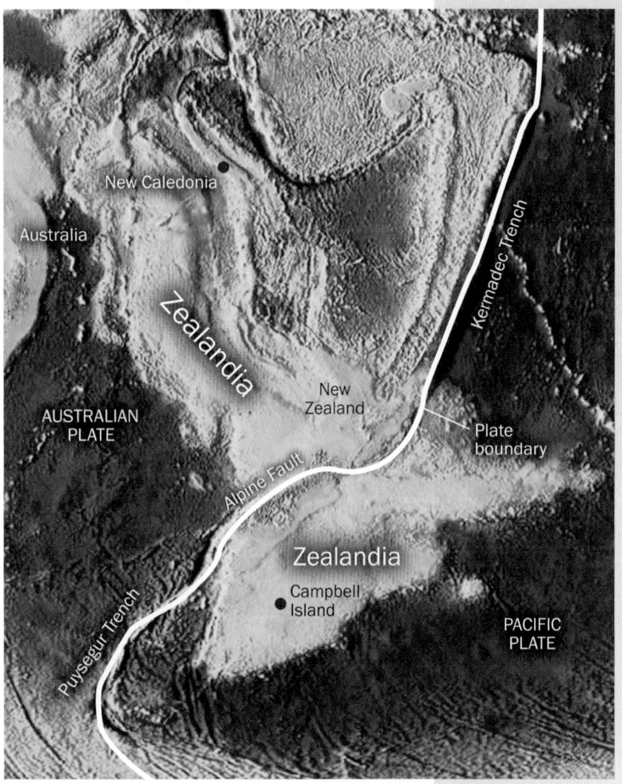

Past mountain building

In geology, the word 'orogeny' is used to describe the process and time of mountain formation by folding and faulting of the earth's crust. The word has Greek origins: *oros* (mountain) + *genesis* (creation) = orogeny (mountain creation).

In New Zealand the processes that resulted in the lifting up of the Southern Alps also caused the formation of the Kaikoura Mountains. This last 25 million years of mountain formation is

Figure 16 The continent of Zealandia. Ninety percent of the continental crust of Zealandia is below sea level (the pale blue shaded area on the map); only small parts of Zealandia, like New Caledonia and New Zealand, are above sea level.

sometimes called the time of the Kaikoura Orogeny, even though the greatest amount of uplift took place along the Southern Alps. The Kaikoura Orogeny is the most recent of mountain-building events (orogenies) that have affected the New Zealand area. The evidence and impact of these older past events has been destroyed by erosion and further mountain building that followed. It is the events of the past 25 million years that gave birth to and shaped the modern-day Southern Alps.

The landforms of the Southern Alps 25 million years in the making — internal tectonic processes have led to mountain formation

Tectonic processes involve huge pressure within the earth's interior causing movement of surface rocks. The shape and landforms of modern New Zealand are the result of such earth processes. The last 5 million years, and especially the last 1 million years, is the time when most of the action has taken place. The Southern Alps are the result of this action.

Between 25 and 15 million years ago the land of modern-day New Zealand was mostly beneath the sea. Tectonic forces in the earth's mantle then became active beneath the crust of Zealandia. This caused an undersea collision between the Australian and Pacific Plates. The Alpine Fault marks the line of the collision (Figure 12). The two colliding crusts have not met head-on but at an angle, in a bit of a sideswipe. This has resulted in a great amount of sideways as well as vertical rock movement. The collision caused compression and crumpling of the crustal rocks so that over time they began to emerge as land (Figures 17 and 18). During the last 5 million years, the Southern Alps have been pushed up to form surface mountains.

Colliding continental crusts

In the area of the South Island, the colliding Australian and Pacific Plates both carry continental crust. When plates carrying continental crust meet, there is no sliding of one plate under the other. Instead, the huge forces put on the crustal rocks by the colliding plates cause the rocks to crumple and snap upwards and downwards in the collision zone. Mountain formation is the result. This means the formation of the Southern Alps has been similar to that of the European Alps and the Himalayas, which are also the result of collision between plates carrying continental crust.

Squeezed up and floating up

There are two parts to the formation of mountainous land in the collision zone.

i   **Folding and faulting processes:** Rocks are placed under great pressure due to the collision, which compresses (squeezes) the rocks. Over time, this causes the rocks to slowly bend (fold) and sometimes to suddenly snap (fault). Upward and downward movement of the rocks takes place and elevated land is the result (Figure 17).

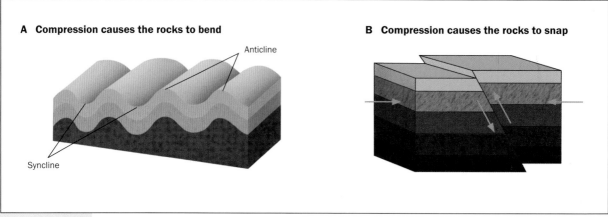

A Compression causes the rocks to bend

Anticline

Syncline

B Compression causes the rocks to snap

Figure 17 Folding and faulting creates mountain landforms.

The crustal rocks of the Australian Plate are stronger than those the Pacific Plate. The Australian Plate has acted like a barrier against which the Pacific Plate has collided, resulting in the rocks of the Pacific Plate being crumpled and forced to rise. This explains why the Southern Alps are higher on the eastern side of the Alpine Fault than they are on the western side, and also why the folding and faulting becomes greater towards the point of collision at the plate boundary. It is a bit like a very slow motion car crash with the front end smashed and buckled while the back end remains undamaged (Figure 18).

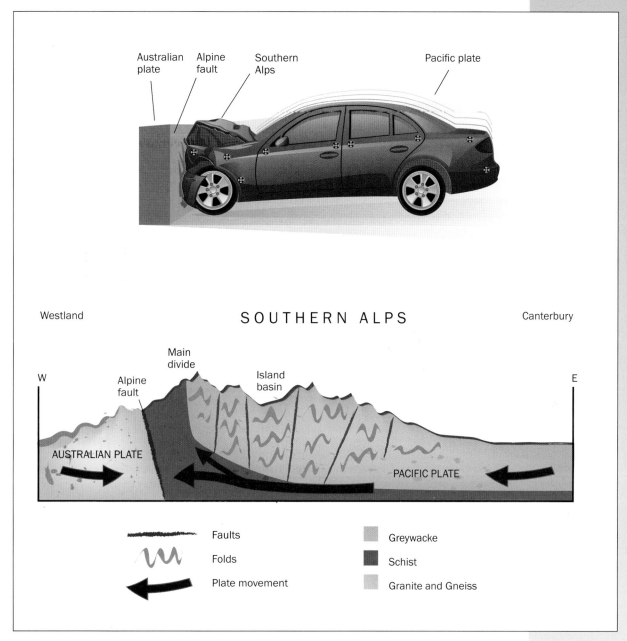

Figure 18 Plate collision has led to mountain formation due to folding and faulting processes.

ii **Buoyancy:** In the collision zone, the rocks have been forced downwards into the interior of the earth as well as upwards. The downward movement of crustal rocks also results in land being elevated. This seems the opposite of what you would expect (counter-intuitive). The reason is because of buoyancy. The continental crustal rocks are lighter and less dense than oceanic crustal rocks. When the plates collide, the continental rocks get crushed and crumpled. They do pile upwards but they get pushed downwards as well. This thickens the continental rocks — they get a deep foundation (root). Because these rocks sit (float) on top of the oceanic rocks, the deeper the foundation, the higher they float. This is like icebergs in the sea where the ice above the sea is supported by deep ice below sea level. The higher the iceberg sits above sea level, the deeper the root it has below sea level (Figure 19 a and b).

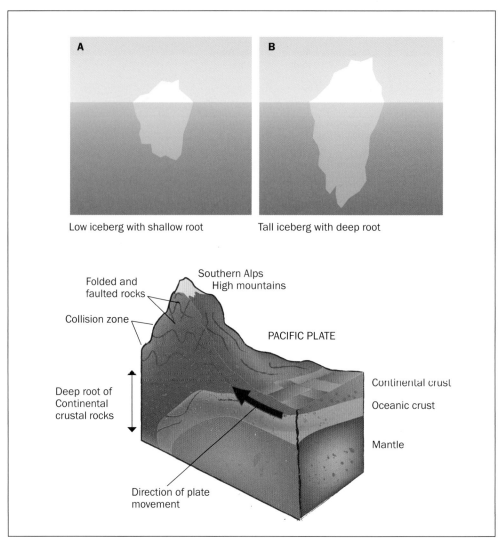

Low iceberg with shallow root

Tall iceberg with deep root

Figure 19 Buoyancy as well as upward folding and faulting have led to the elevation of the Southern Alps.

The processes continue today

The speed of the plate collision, and resulting land movement and mountain building, has accelerated during the past 25 million years and has been especially rapid in the last 5 million years. The compression and upward and sideways movements are still going on. The mountains today are rising as fast as at any time in the past. In human terms, an annual speed of 3 centimetres of horizontal movement and 1 centimetre of vertical movement seems slow (the speed of fingernail growth), but on a geological timescale this is fast movement.

Away from the collision zone (towards the eastern foothills of the Alps and Canterbury Plains), land movement has been less. The Southern Alps have been the focus of the crash zone and of the rock crumpling.

Highest and largest outcomes of tectonic processes

New Zealand's largest mountain range, the Southern Alps, and New Zealand's highest mountain peak, Aoraki/Mount Cook, are the result of the collision zone tectonic processes. These same tectonic processes have also led to the formation of New Zealand's largest intermontane (lower land surrounded by mountains) basins. The Mackenzie Basin is the largest of all, covering an area 30 km by 50 km. It has been left as an area of lower land surrounded by mountains, as land around it has been faulted and folded upwards.

Learning Activities

1   Match up the two columns of information.

Column A	Column B
Greywacke	A period of time when land moves and mountains are formed
Schist	New Zealand's largest intermontane basin: lower and flatter land surrounded by mountains
Zealandia	Name of the great fault line that runs through and along the Southern Alps
Orogeny	The plate on which the eastern side of the Southern Alps and High Country are located
Tectonic	To do with floating and upward forces — how mountains are like icebergs
Alpine	Forces and processes that bring about changes to the shape of the crust of the earth
Compression	A shiny and layered metamorphic rock; greywacke changed by pressure and heat
Pacific	New Zealand's own continent that is mostly hidden beneath the ocean
Buoyant	Squeezing and pressure put on rocks that can cause folding and faulting
Mackenzie	A hard grey-coloured sedimentary rock that is found all across the Southern Alps and High Country

2   Give two examples of spatial variation shown in Figure 18.

3   A process involves a sequence of actions or events that shape and change places and the environment. Copy and complete the following diagram to show how internal tectonic events lead to the formation of mountains. Choose from the ideas i–vi. Give your completed diagram a title.

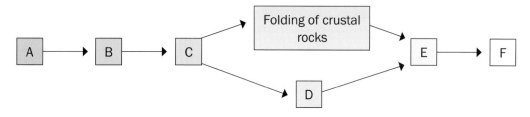

i   Faulting of crustal rocks

ii   Heat, pressure and convection movement in the mantle beneath the earth's crust

iii   Mountains and basins are formed

iv   Surface rocks are distorted and move upwards and downwards

v   Crustal stresses and plate movement

vi   Plate collision and rock compression

4   Write paragraphs explaining how internal tectonic forces and processes have led to the formation of Southern Alps and High Country landforms. Include these four things within your answer:

i   reference to plates, rock types and to folding and faulting

ii   explanation of the part played by buoyancy

iii   a time dimension (when things took place)

iv   at least one diagram.

C: Battleground

Important understanding: The landforms of the Southern Alps and High Country are the outcome of the operation of both internal and external natural processes.

Almost as fast as the land has been pushed up, weathering and erosion have worn away the land. This weathered and eroded material has been transported away from the mountains by glaciers and rivers. The eroded material has been deposited in mountain valleys and basins, across lower land and also out at sea. Without this weathering and erosion, the Southern Alps would be 20,000 metres high — this is six times higher today than they actually are.

MOUNTAINS CREATED ELEVATED LAND

Buoyancy
Folding processes
Faulting processes
TECTONIC AND OTHER INTERNAL PROCESSES

EXTERNAL PROCESSES
Weathering
Mass movement
Glacial processes
River processes

MOUNTAINS WORN DOWN AND LANDSCAPE RESHAPED

Figure 20 Landforms and the battle between internal and external processes.

Learning Activities

1 In a brief sentence, describe the landforms shown in the photo in Figure 20.
2 What processes have created these landforms?

D: The fall of the Southern Alps — external surface processes at work

Important understanding: The most recent period of global glaciation began 75,000 years ago. This and other surface processes operating since that time have reshaped the landforms of the Southern Alps and High Country.

The last 75,000 years have been especially important in producing the landforms that make up the present-day Southern Alps environment — tectonic forces have set the scene over millions of years but surface processes over the last 75,000 years have modified and fine-tuned these landforms.

Weathering

Weathering is the name given to processes connected with the action of the weather on surface rocks that causes them to disintegrate and break up. In mountain areas like the Southern Alps, the most important type of weathering is called mechanical or physical weathering. This type of weathering involves:

i **Expansion and contraction** of rocks because of changes in temperature. The sun heats the rocks during the day, with cooling of the rocks then taking place at night. The heating causes the rocks to expand, while at night contraction of the rocks takes place as they cool down. Over time, this process causes small cracks in the rocks to develop.

ii **Freeze-thaw**: Water from rain, and from snow and ice melt, enters cracks caused by expansion and contraction, and cracks resulting from folding and faulting processes. In the Southern Alps, the cold night-time temperatures frequently result in this water freezing in the cracks. This puts pressure on the rocks because water expands as it freezes. The pressure is released during the day when the ice melts. This repeated freeze-thaw process speeds up disintegration of the rocks.

A large amount of loose surface rock results from weathering.

Figure 21 Weathered rocks and scree slopes high on mountain hillsides above Arthur's Pass.

Mass movement — rock and soil avalanches

Much of the land in the Southern Alps consists of steep slopes, and on such slopes **gravity** causes the loose surface rocks to tumble and slide downhill. Often this happens suddenly and without warning. A process called '**frost heave**' adds to this downhill movement in a regular way. When temperatures drop below freezing point, ice crystals form in the surface rock and soil layer. The ice crystals grow in a vertical direction and lift up soil and rock fragments as they grow. When the crystals melt, the soil and fragments drop downslope. Sudden downhill movements of rock take place when the fractured and loose rock becomes unstable, or when earthquakes cause mountains and hills to collapse.

Rock avalanche

The rock avalanche on Aoraki/Mount Cook in 1991 (page 10) took place because of the weak and fragmented rock giving way due to gravity — this was a natural process waiting to happen. Events like this take place frequently and usually without warning within this mountain environment. All of this soil and rock moves towards the foot of the slopes where it builds up deep layers of loose rock and soil. This loose layer is called a **scree** deposit. Scree slopes are a common landform feature found in the Southern Alps (Figure 21 and 23). Like earthquakes, other irregular but extreme natural events also cause lots of erosion in mountain areas. After heavy rain, massive amounts of water run off the land, and hillside collapses are common. This leads to **debris flows** where huge amounts of loose mud, gravel and boulders flow rapidly downhill mixed up with water like wet concrete. Landforms called alluvial fans fill many lower valleys in the Alps where the debris flow material has been deposited (Figure 22).

Figure 22 Rockfall and debris flow.

Figure 23 Otira road viaduct — scree accumulation in the valley of the Otira River.

E: Glaciation

A sequence of ice ages followed by warmer periods (called interglacials) have affected the world over the last 2.5 million years; there have been 20 glacial and interglacial cycles during this time.

The most recent of these ice ages (called the Otira Glaciation in New Zealand) began about 75,000 years ago and only ended when temperatures began warming 14,000 years ago. Since then, temperatures have risen to their present level and the world is now in the middle of another interglacial period (Figure 24).

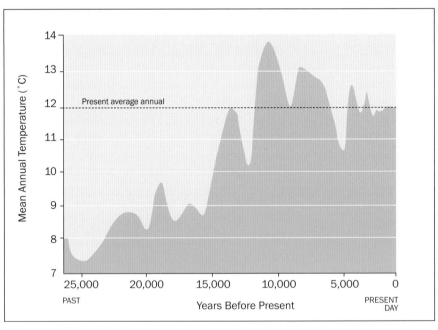

Figure 24 New Zealand temperature changes — end of the Otira Glaciation and the present-day interglacial.

Little evidence of the earliest glacial and interglacial periods exists but glaciation has had a massive influence on present-day landforms of the Southern Alps and High Country area. The most recent Otira Glaciation destroyed and reshaped the earlier landforms. During the Otira glacial period, New Zealand temperatures were 4–5°C lower than they are today; the sea level was also lower and the North and South Islands were joined (Figure 11). In the Southern Alps and High Country, the permanent snowline was much lower than it is today. Ice and snow covered the lower mountain peaks, and huge glaciers filled the river valleys and spread out from the mountains into lowland areas.

ISBN: 9780170233316

Glacial processes

Glacial processes have created landforms in the Southern Alps and High Country that make them different from those in other parts of New Zealand. This is the reason the environment is described as a 'glaciated environment'. Ice has shaped and reshaped the landforms — it has given the fine detail to the landforms created by tectonic forces (Figure 25).

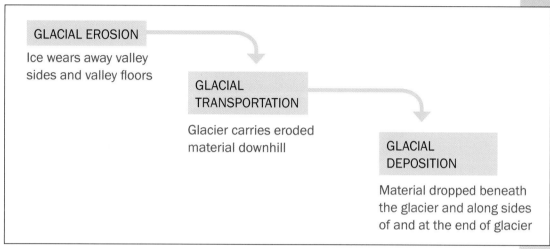

Figure 25 Three interrelated glacial processes.

How the glacial process works

Glacial erosion

Valley glaciers are powerful agents of erosion. They are like slow-moving rivers of ice. They reshape the land as they travel downhill from their source areas high in mountains to their terminus (snout) in warmer areas lower down. Glaciers wear away and smooth the sides and floors of valleys as they move down valleys by two actions:

i   **Abrasion**. As result of mass movement on hillsides above a glacier, rock fragments fall onto the surface of the glacier. As well as remaining on the glacier surface, some of this rock works its way into the sides and base of the glacier. As the glacier moves, these rocks carried within the glacier scrape and grind away the sides and floor of the valley.

ii **Plucking**. Glacial ice works its way into and around cracked and loosened rock found along the sides and floor of a valley. As the glacier moves downhill, it pulls (plucks) away these cracked rocks and loosened rocks it has frozen onto.

Glacial transportation

These eroded rocks are carried within the glacier as it continues flowing downhill. The moving glacier also carries downhill any rocks that remain on its surface. Glaciers transport huge amounts of material. The material carried by a glacier is called 'till'. Most till is sharp and angular in shape and varies in size from that of a small car down to tiny pieces of rock.

Glacial deposition

As a glacier travels downhill and moves into lower and warmer areas, melting of the glacier ice takes place. The rocks, boulders and gravels (till) transported by the glacier get deposited beneath the melting ice and along the sides and at the end of the melting glacier. New landforms, called moraines, are created by these glacial deposits.

Many distinctive landform features are created by glacial processes

The result of both glacial erosion and glacial deposition is the formation of distinctive landforms features in all glaciated mountain areas (Figure 26). Because the features were studied first in Europe, many of the names used to identify them are of French, German and Scandinavian origin.

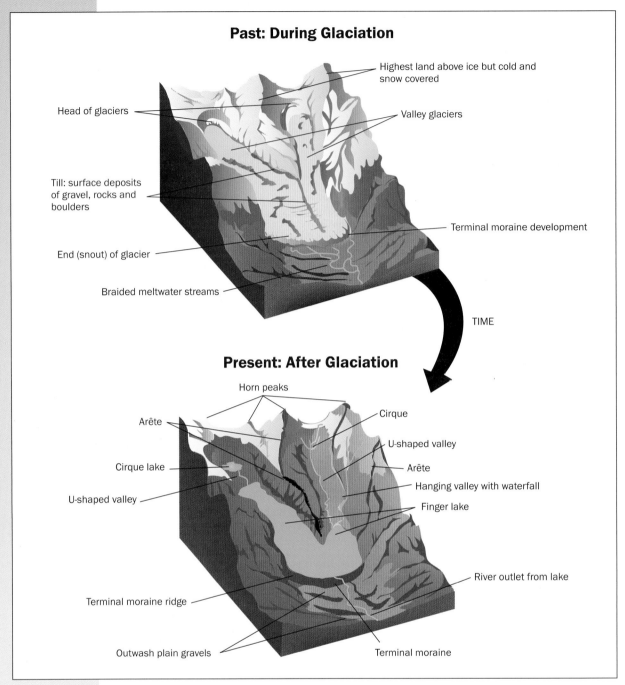

Past: During Glaciation

Highest land above ice but cold and snow covered

Head of glaciers

Valley glaciers

Till: surface deposits of gravel, rocks and boulders

Terminal moraine development

End (snout) of glacier

Braided meltwater streams

TIME

Present: After Glaciation

Horn peaks

Arête

Cirque

Cirque lake

U-shaped valley

U-shaped valley

Arête

Hanging valley with waterfall

Finger lake

River outlet from lake

Terminal moraine ridge

Outwash plain gravels

Terminal moraine

Figure 26 Glacial landforms.

- U-shaped valleys are the result of glacial erosion and are found all through the Southern Alps. Those on the western side of the Main Divide like those in the Franz Josef and Fox valleys are short and steep. Also on the western side are the U-shaped valleys cut into lower land by past glaciers that have since been flooded and drowned by sea water to form coastal inlets. These 'glaciated inlets' are called fiords. Milford Sound is the most famous of these but the whole Fiordland coast has such inlets. On the eastern side of the Alps, the U-shaped valleys like the Tasman and Dart (Case Study 1, page 31) are longer and more gently sloping than the western side ones.
- Often the big glaciers in the main valleys have eroded downwards more quickly than the smaller tributary (side) glaciers have done. After the glaciers have melted and retreated, the side valleys no longer join the main valley in a smooth way. Instead, they have been left high above the level of the main valley. These side valleys are called hanging valleys. Rivers from these hanging valleys now flow out via waterfalls into the main valley below.

 ISBN: 9780170233316

- Where glaciers eroded rock from both sides of ridges that separate two valleys, the ridge becomes narrower and narrower and ends up as a sharp ridge called an **arête**. Aoraki/Mount Cook and Mount Aspiring both have arêtes that lead up towards their summits (Figure 6).
- Cirques are bowl-shaped basins cut into the rocky slopes by ice erosion in the high elevated areas where glaciers begin. The basins can be as large as 2 km across. When the glaciers disappear, the basins are often filled with loose rock because of rock falls from the weak schist rocks on the steep slopes above them. In some locations, the basins become filled with water forming distinctive circular-shaped lakes within the mountains. Cirques are found all through the higher parts of the Southern Alps (the Dart Glacier and Lake Wakatipu Case Study on page 31 has an example in Lake Alta).
- High above the cirques and arêtes are mountain tops with distinctive horn (pyramid) peaks. Mount Aspiring and Aoraki/Mount Cook (Figure 6) are both examples. Glacial erosion and weathering taking place on all the sides of the mountain result in the formation of such peaks.
- Where glaciers flow over small rocky hills, they smooth off the side of the hill they flow over first due to abrasion, but on the downstream slope the glacier ice freezes onto the rock and this side becomes rough and ragged due to the plucking action of the ice. Hills shaped this way are called roches moutonnées (the Dart Glacier and Lake Wakatipu Case Study on page 31 has an example).
- On lower land around towards and at the end of the glaciers, lots of the eroded boulders, rocks and gravels get deposited. Material deposited where the glacier ends can build up to form a ridge. This ridge is called a terminal moraine, while deposits left along the side of the glacier are called lateral moraines.
- In many of the U-shaped valleys to the east of the Main Divide, ice meltwater and river water has become trapped by terminal moraine barriers (natural dams) resulting in the formation of deep, long and narrow lakes like Te Anau, Pukaki and Hawea. Because of their shape, these lakes are called finger or ribbon lakes.

Rivers in the mountains

Rivers are another natural agent involved in the shaping of the mountain area landforms. Just like glaciers, mountain streams and rivers carry out the processes of erosion, transportation and deposition. These three river processes are called fluvial processes. In the past, rivers eroded V-shaped valleys as they flowed away from their sources high in the uplifted mountains towards the sea. When glaciers moved down these valleys during times of colder climates, they further eroded these valleys both sideways and downwards to create distinctive U shapes. Since the last glacial advances, rivers have reappeared in the U-shaped glaciated valleys and have begun reworking the deposits left by the glaciers. Meltwater flowing from beneath and from the end of a glacier is often the starting point for these rivers. Some rivers are now cutting new V-shaped valleys into the moraines deposited by glaciers across floors of the wide U-shaped glaciated valleys. Many rivers carry away lots of the material deposited by the glaciers. The further the material is carried by rivers, the smoother it becomes as it bounces and rolls along the bed of the river. The material also becomes smaller, with lots of fine sediment produced by the river-transporting process. This material then gets deposited by the rivers across land well downstream of the end of the glacier. These deposits are called 'outwash plains'. Lower valleys of rivers like the Tasman are covered in outwash material. Beyond the mountains, rivers flowing west like the Haast River and east like the Rakaia River have transported and then deposited the ground-up glacial material across lowlands, helping to build up the Haast coastal lowlands and the Canterbury Plains.

Figure 27 Rob Roy Stream fed by ice melt from the Rob Roy Glacier in Mount Aspiring National Park.

Learning Activities

1   Match up the heads and tails statements from columns A and B, respectively, to make 10 sentences about landforms of the Southern Alps and High Country.

Column A — Heads	Column B — Tails
a The Otira Glaciation is the name of the	surface rocks and the formation of scree slopes.
b Without weathering and erosion wearing down the	of the process of glaciation and create new landforms.
c Weathering is a process that results in the break-up of	often dam rivers and cause ribbon lakes to form.
d Rock avalanches on steep and unstable slopes	valleys on the coast that have been flooded by the sea.
e Glacial erosion and glacial deposition are both parts	material that was originally produced by glacial erosion.
f Glaciers carry lots of material downhill	most recent ice age to affect New Zealand.
g Fiords like Milford Sound are glacially eroded U-shaped	within the ice and on the surface of the ice.
h Aoraki/Mount Cook is an example of a pyramid-shaped	often happen suddenly and without warning.
i Terminal moraines are glacial deposition features that	land the Southern Alps would be 20,000 metres high.
j Outwash plains are made up of river deposited	peak formed by glacial erosion and weathering.

2   Refer to the information about weathering and mass movement on page 25 and Figures 21 and 23. Either describe the characteristics (main features) of scree slopes and explain how they are formed, or draw a fully annotated sketch of Figure 21.

3   a   Make a copy of Figure 24. On the graph, write in their correct places 'Otira Glaciation' and 'Present-day interglacial'.

   b   Refer to the information about glaciation shown in Figures 25 and 26 and text information on pages 26–29.

      i   Give an example of one landform that results from glacial erosion and one that results from glacial deposition. Draw diagrams/sketches of each landform you choose.

      ii   'Glaciation' is an example of a natural process. In an illustrated essay explain how this process operates (remember to include information about sequence of events and the effects the process has on landforms).

4   Explain two ways rivers have shaped and reshaped landforms in the Southern Alps and High Country area.

5   **Summary: How landforms of the Southern Alps and High Country have developed and changed over time.**

   Make a copy of the flow diagram and give it a title. Copy the table but correctly match the events with the time period they took place in. (**C** has been matched as an example.)

PAST   **A** ⟶ **B** ⟶ **C** ⟶ **D** ⟶ **E** ⟶ **PRESENT**

	Date — Time Period	Jumbled events
A	200–300 million years ago	Pacific and Australian Plate collision; modern Southern Alps formation begins with folding and faulting of crustal rocks **(C)**
B	25 million years ago	Greywacke being formed under the sea; an old sedimentary rock
C	5 million years ago	A battleground — uplift of land continues but rapid erosion due to glaciers, weathering, mass movement and rivers continually wears away and reshapes the land
D	75,000–14,000 years ago	Kaikoura mountain-building orogeny: land begins to be pushed up from beneath the sea
E	14,000 years ago – present	Most recent ice age; widespread glaciation across the higher parts of the South Island creates a range of glacial erosion and glacial deposition landforms

Case Study 1

The Dart Glacier, Dart River and Lake Wakatipu

Important understanding: The glacier, the river and the lake are all part of the same landform system.

Wakatipu is New Zealand's longest lake (80 km) and third largest in area (Taupo and Te Anau are larger). The lake is over 400 metres deep and the floor of the lake is below sea level. Viewed from above, the lake has a distinctive 'snake-like' shape (Figure 28).

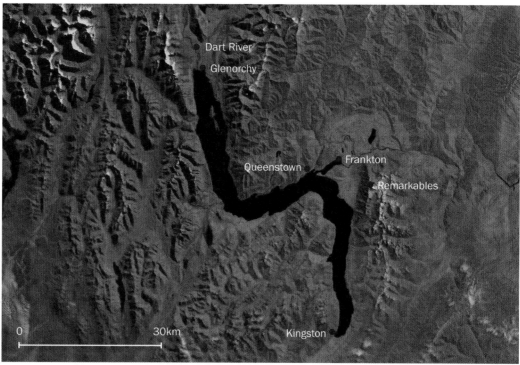

Figure 28 Satellite image of Lake Wakatipu, a finger lake in a U-shaped valley surrounded by high glaciated mountains.

The origin of Lake Wakatipu

Lake Wakatipu and the mountains — Maori legend
Ngai Tahu legends tell the story of the origin of the lake.

Long ago there was a chief who had a beautiful daughter called Manata. Many young men wanted to marry her, including a young warrior Matakauri, who was Manata's sweetheart. But the chief, Manata's father, would not let the couple marry as he thought Matakauri was not good enough for her.

One day a terrifying taniwha (giant) named Matau came and stole Manata. The chief was heartbroken and said that anyone who rescued his daughter could marry her. Even the strongest of the young warriors were frightened at the thought of fighting the taniwha. Matakauri's love for Manata was so strong, however, that he set out in search of the taniwha. After finding Matau in the mountains he observed that whenever a nor'west wind blew, the taniwha went to sleep. While Matau slept, Matakauri crept close and found Manata but she was tied to the taniwha with strong ropes. He tried to cut the ropes but he failed. Manata sobbed bitterly and so great was her love for Matakauri that when her tears fell on the ropes, the love in her tears dissolved them. The pair escaped and ran away together. Manata's father, kept his promise and allowed the couple to marry.

Matau, however, still lived in the mountains and Matakauri decided to deal with him once and for all. He waited until there was a strong nor'wester blowing and the taniwha was asleep. He then set light to the bracken bed that Matau was sleeping on. Matau was surrounded by flames and the fat from his burning body made the flames even hotter and stronger, burning a hole deep in the ground. The fire also melted the snow on the hills and water poured in filling up the hole left by Matau's burnt body.

The outline of Lake Wakatipu resembles the shape of a person lying down, the shape of the body of a giant: the head at Glenorchy, knees bent at Queenstown and feet at Kingston.

For Ngai Tahu, traditions and stories such as this provide links between the world of the gods, the past and present generations. These stories reinforce tribal identity and tell of the events which shaped the environment of Te Wai Pounamu, the home of Ngai Tahu.

Lake Wakatipu and the mountains — geologist's account

Today, water from the Dart River flows into the northern end of Lake Wakatipu at Glenorchy. The river begins life as meltwater flowing from the end of the Dart Glacier high in the Mount Aspiring National Park. The glacier originates at an even higher elevation of over 2000 metres in snowfields close to the Main Divide. The glacier travels downhill at a speed of about half a metre per day — about 200 metres each year.

The distance from Kingston at the southern end of Lake Wakatipu to Glenorchy at its northern end, and then up the Dart Valley to the top of the Dart Glacier, is over 140 kilometres. The glacier, the river and the lake are all part of one connected 'water' system (Figures 29–35).

Eighteen thousand years ago things were very different. The world was in the grip of the most recent great ice age. At this time the Dart Glacier extended all the way down from the high mountains to Kingston. Lake Wakatipu then was not a freshwater lake but a deep valley full of glacier ice — this was the lower part of the Dart Glacier. Since this time the earth has warmed and is now in an interglacial period. Much of the ice has now melted and the Dart Glacier has retreated to be 135 kilometres away from Kingston.

There have been three other great ice ages and glacier advances before the one of 18,000 years ago. Each time the Wakatipu and Dart valleys have been filled by huge glaciers. These glaciers have eroded the valleys formed originally by rivers and created distinctive U-shaped glacial valleys with wide flat bottoms and steep sides. Lake Wakatipu is located in one of these valleys, and is constantly filled with water melting from glacial ice and snow, plus rain and river water.

At Kingston there is a terminal moraine marking the point where the glacier ended 18,000 years ago. This moraine forms a natural dam at the southern end of the lake. Today, water from the lake no longer flows through Kingston and into the Mataura River as it used to, but instead exits the lake at Frankton and flows into the Kawarau River. South of Kingston there are other older terminal moraines that mark the end point of glaciers from earlier ice ages when the glaciers extended even further away from the mountains.

Meltwater streams and rivers flowing away from the end of the glacier have transported and redistributed material once deposited by the glacier. These river deposits are called outwash deposits. Those below Kingston are old deposits from the ice ages of the past, while today in the valley of the Dart River between the end of the glacier and Lake Wakatipu, the deposits are still being added to.

Figure 29 Looking north. The northern part of Lake Wakatipu with the Dart Valley and the mountains of the Mount Aspiring National Park and Main Divide in the background.

Case Study 1

Figures 30–36 From top to bottom — Dart Glacier, Dart River and across Lake Wakatipu

Figure 30 The source area and upper part of the Dart Glacier in the Mount Aspiring National Park. The till (dark material) on the surface of the glacier is made up of boulders, rock and gravel that have fallen onto the glacier from the weathered sidewalls of the valley.

Figure 31 Dart River. Braided river pattern and outwash deposits in the U-shaped valley to the north of Glenorchy.

Case Study 1

Remarkables

Peninsula Hill –
Roche Moutonnee

Lake Wakatipu – finger lake
in a U shaped valley

To Frankton and
Kawarau River

Queenstown Bay

Figure 32 Lake Wakatipu looking south from Queenstown towards Kingston.

Figure 33 Lake Alta, a cirque lake cut into the schist rock of The Remarkables.

 ISBN: 9780170233316

Case Study 1

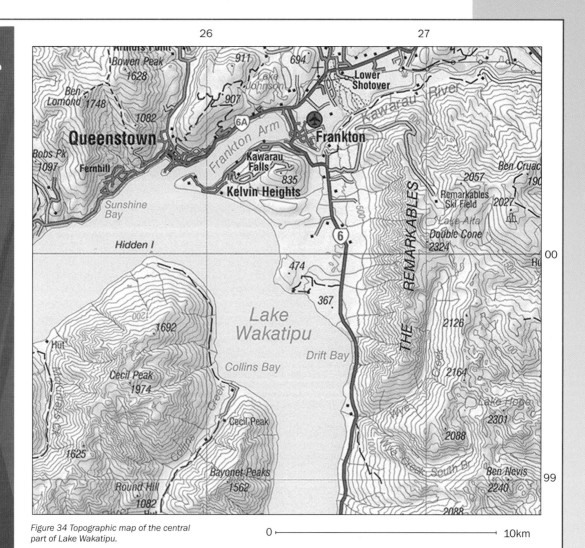

Figure 34 Topographic map of the central part of Lake Wakatipu.

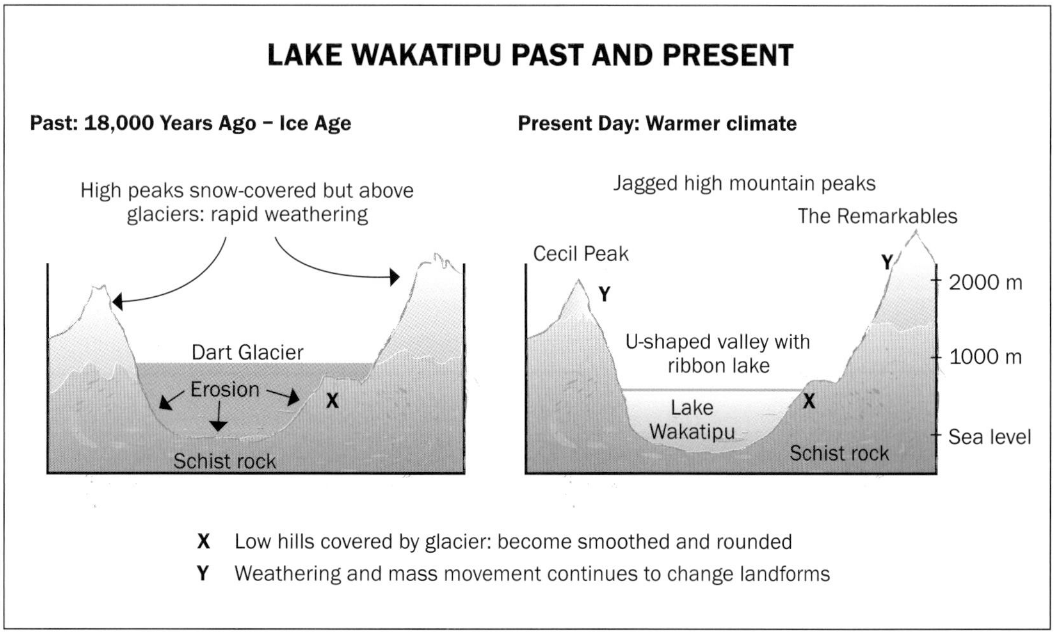

0 ⊢————————————————⊣ 10km

LAKE WAKATIPU PAST AND PRESENT

Past: 18,000 Years Ago – Ice Age

High peaks snow-covered but above glaciers: rapid weathering

Dart Glacier

Erosion → X

Schist rock

Present Day: Warmer climate

Jagged high mountain peaks

The Remarkables

Cecil Peak

Y

Y

2000 m

U-shaped valley with ribbon lake

1000 m

Lake Wakatipu

X

Schist rock

Sea level

X   Low hills covered by glacier: become smoothed and rounded

Y   Weathering and mass movement continues to change landforms

Figure 35 Lake Wakatipu during and after glacial advances.

Case Study 1

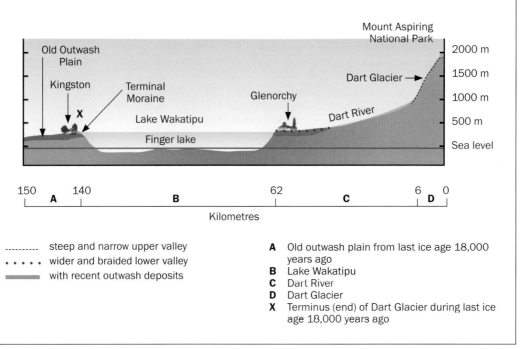

LAKE WAKATIPU AND THE DART VALLEY TODAY LONG PROFILE

---------- steep and narrow upper valley

• • • • • wider and braided lower valley

▨▨▨▨ with recent outwash deposits

A Old outwash plain from last ice age 18,000 years ago
B Lake Wakatipu
C Dart River
D Dart Glacier
X Terminus (end) of Dart Glacier during last ice age 18,000 years ago

Figure 36 The Dart Valley and Lake Wakatipu.

Learning Activities

1 **a** How is the origin of Lake Wakatipu explained in the Maori legend?

  b How do geologists explain the origin of the lake?

2 Refer to Figures 28 and 34.

  a Describe the shape of Lake Wakatipu.

  b In which direction does the Dart River flow?

  c Calculate the distance between Glenorchy and Kingston in a direct line.

  d Describe the location of Queenstown.

  e What is the altitude and name of the highest point in the area shown on the topographic map?

  f Name the peak located at 257994.

  g Give the six-figure grid reference of the cirque Lake Alta.

  h How wide is Lake Wakatipu between the shoreline at Collins Bay and the shoreline at Drift Bay?

  i Describe the location of The Remarkables Ski Field.

  j There is a lot of steep land in the area shown on the topographic map. How can you tell where the steep land is located from reading the map?

3 Refer to the photos in Figures 30–33 and the diagrams in Figures 35 and 36. Describe how landforms change from the head of the Dart Glacier to the outwash plain south of Kingston at the southern end of Lake Wakatipu. Include specific detail and a diagram as part of your answer.

  ISBN: 9780170233316

Vegetation of the Southern Alps and High Country

Important understanding: Vegetation varies from west to east across the Southern Alps and High Country and also with height above sea level.

Like landforms, the vegetation of the Southern Alps and High Country is distinctive. Yellow and golden tussock grassland and dark-green beech forests are the two most widespread types of vegetation found within the environment.

Many factors influence vegetation in the Southern Alps and High Country area. These factors are shown in Figure 37. The most important of these factors is height above sea level (altitude), which in turn controls temperatures and the amount of rainfall.

Below the treeline, beech forest and tussock grasslands are the main vegetation types. Above the treeline, vegetation changes quickly as both land height and slope angle increase. A zone of low shrubs quickly gives way to slopes covered with tall tussocks and herb fields. Higher still there is bare rocky ground with snow and ice covering the highest slopes and mountain tops.

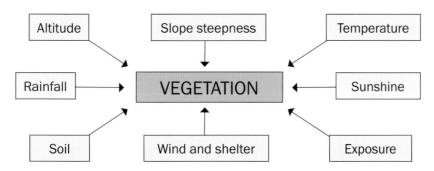

Figure 37 Elements (factors) that influence vegetation.

Conditions vegetation must adapt to:
Life for plants is difficult in the steep, high, glaciated, unstable and stormswept landscape of the Southern Alps and High Country. Plants face problems of:
- winter temperatures that are below freezing with snow and ice covering the land
- high daytime temperatures in summer with rapid cooling at night often to below 0°C
- strong and persistent winds
- areas with very low rainfall but high evaporation rates, which dries out plants in summer
- thin and infertile soils on the steep slopes
- unstable soils on scree slopes that are always moving.

Plants have adapted to these harsh conditions by:
- Having wiry, tough branches to reduce wind damage; having closely packed leaves that are small and tough, which makes them less easily frozen or dried out and less likely to be damaged by wind, hail and snow; having leaves with hairs on them to reduce air movement over the leaf surface and thereby protect the leaves from the cold.
- Having a deep root system to provide strong anchorage and to tap into the water and nutrients that are available deep below the surface in the mountains.

- Growing in closely packed clusters so they can trap warm air and moisture and protect themselves from wind, ice penetration and movements of snow down a slope.
- Having large white and yellow flowers — this is because the plants rely on pollination by flies, moths, small bees and beetles, which are attracted more to these two colours than to other bright colours.

Vegetation — spatial variation

A close-up view shows that within a small area, differences in vegetation can occur because of variations in land steepness, in soil type and quality and in exposure to rain, wind and sun.

An overview of the whole Southern Alps and High Country area, on the other hand, shows two much bigger changes in vegetation:

i **Variation from west to east across the Southern Alps and High Country.** This is due mainly to rainfall amounts decreasing eastwards. On wetter slopes west of the Main Divide, temperate rainforest with species like beech, kahikatea and rata is the main vegetation type, while at the same altitude on the drier east side of the Main Divide, tussock grassland dominates.

ii **Vertical zonation.** This is where vegetation type changes with altitude, resulting in zones of vegetation up the mountainsides (Figure 38). Vegetation in the valleys is very different from that found on the higher slopes and, in turn, different vegetation, or no vegetation at all, is a feature of the highest slopes and mountain peaks.

In the highest areas above the treeline, vegetation is natural — it is adapted to the natural conditions. In the lower areas below the treeline, the influence of people is significant — beech forest and tussock grasslands have both been influenced by the actions of people. Burning of the vegetation by Maori and Europeans, both accidentally and deliberately, has altered the natural vegetation patterns. There are fewer trees and a greater amount of tussock than there would have been without human interference.

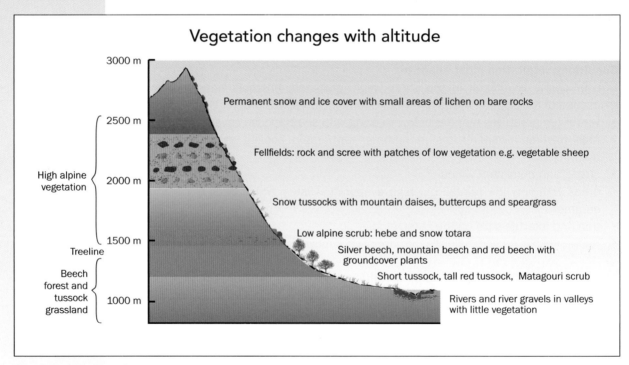

Figure 38 Vertical zonation of vegetation.

Figure 39 Highest peaks — bare rock and permanent snow and ice cover.

Figure 40 River valley with beech forest leading up to the treeline.

Figure 41a and b Fellfield scree and vegetable sheep vegetation.

Figure 42 Snow tussock, speargrass and mountain daisies.

ISBN: 9780170233316

Learning Activities

1 Explain the meaning of each of these terms:

   a   treeline

   b   tussock

   c   vegetation vertical zonation.

2 Give two examples of problems faced by vegetation growing in the Southern Alps and High Country environment and examples of how the vegetation has adapted to the conditions.

3 a   Give an example of spatial variation in vegetation within the Southern Alps and High Country environment.

   b   Explain why this spatial variation occurs.

4 *'The vegetation of the Southern Alps and High Country may look as if it is all natural, but in fact it is not.'* Explain this statement.

Soils are generally of low quality

Important understanding: Soils of the Southern Alps and High Country are mostly of poor quality.

Across a lot of the Southern Alps and High Country, especially in the higher and steeper areas, there is no soil at all, only large slabs of bare rock. Many hillsides are covered in scree — masses of loose pieces of rock (shingle). Where soils have formed, they are mostly thin and of low fertility. The rocks like greywacke, from which the soils are formed, are low in nutrients. In addition, water has removed minerals and nutrients from the top layers of soil and rivers have then carried these minerals away from the mountains in suspension and solution. This process is called leaching.

In some rock crevices on the higher slopes, fine sediment deposits have built up and plants have become established in such places. Elsewhere on more gentle slopes, weathering has broken up the rock to form soil, and some wider valley floors and basins have had a build-up of finer sediment deposits from glaciers (moraines) and rivers. Once vegetation gets established, leaf and other plant material from the trees, bushes and grasses slowly add organic matter to the soil. Soils of higher quality in these locations are the result. However, stony, poor-quality leached soils like those of the Mackenzie Basin remain the soil type of most of the lower areas. Seventy percent of Mackenzie Basin soils are outwash deposits that are shallow and stony with low fertility. They fall into the two lowest-grade soil classes, classes 4 and 5.

Learning Activity

1 Copy and complete this star diagram to show four features of the soils of the Southern Alps and High Country region.

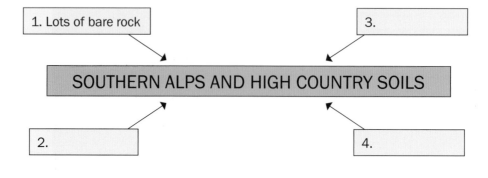

1. Lots of bare rock

3.

SOUTHERN ALPS AND HIGH COUNTRY SOILS

2.

4.

   ISBN: 9780170233316

Climate of the Southern Alps and High Country environment

Important understanding: A distinctive feature of the climate of the Southern Alps and High Country environment is the great variation that exists within the region. Climate varies over short distances; weather (day-by-day and week-by-week conditions) can vary from place to place but can also be unpredictable and change rapidly (Figure 43).

Climate features of the Southern Alps and High Country, an area with climate extremes and climate variety, include the following:

- There are four climate types within this one environment area (Figure 44).
- Strong and persistent west and northwest winds prevail across all of the Southern Alps and High Country. On the mountaintops, winds of over 200 kph have been recorded.
- There is rainshadow effect (see pages 254–255 and Figure 35 in the Global Mountains chapter) that is one of the most remarkable in the world. Over a distance of only 50 km across the Southern Alps, rainfall decreases from over 8000 mm per year on the western side of the Alps to less than 800 mm per year on the eastern side.
- The 'nor'wester' (fohn) effect influences the area. Northwest winds leave the Tasman Sea full of moisture and then rise over the barrier of the Southern Alps. As the air rises, it cools, forms clouds and produces heavy falls of rain and snow. Most of the rain and snow falls to the west of the Main Divide, but some spills over onto high areas east of the Divide. Once away from the Main Divide, the air has lost most of its moisture and gathers strength as it descends. The northwest winds blast into eastern valleys and inland basins of the High Country as warm and dry gale-force winds, which frequently raise dust clouds from the bare moraines and riverbeds.
- Sheltered inland basin areas have high summer temperatures but in winter have some of the coldest temperatures of anywhere in New Zealand.
- Air masses from the south (from the Antarctic) — 'polar blasts' — bring days of freezing temperatures and heavy dumps of snow across the lands of the eastern High Country during winter. Roads become blocked, electricity supplies disrupted and farming and tourism both face huge challenges in such conditions.
- Central Otago and the Mackenzie Basin have annual sunshine hours that are as high as anywhere in New Zealand, but Fiordland holds records for the greatest number of cloudy days each year.
- In the eastern High Country and inland basins, the difference between the highest daytime temperatures and the lowest night-time temperatures (the diurnal temperature range) is huge.
- The weather (day-to-day conditions) changes frequently and quickly.

Figure 43 Milford Track – weather changes rapidly in the mountains.

- The ocean has a big influence of the climate west of the Main Divide, which is wet, with mild winters and warm summers. This is called a maritime effect.
- East of the Main Divide in Central Otago and in inland basins like the Mackenzie Basin, the influence of the ocean on climate is limited. This results in a climate that is semi-continental — dry and sunny with hot summers and very cold winters.
- There are many places with a microclimate. A microclimate is the name used to describe a small area with a climate that is different from surrounding areas. For example, the valleys in the Southern Alps and High Country hillsides that face north get much more sun than the south-facing hillsides. The north-facing hillsides can be 5°C warmer than the south-facing ones.

A

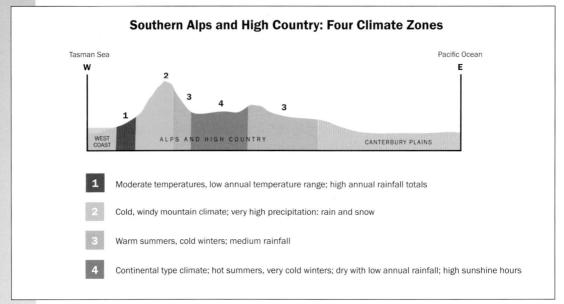

Southern Alps and High Country: Four Climate Zones

1 Moderate temperatures, low annual temperature range; high annual rainfall totals

2 Cold, windy mountain climate; very high precipitation: rain and snow

3 Warm summers, cold winters; medium rainfall

4 Continental type climate; hot summers, very cold winters; dry with low annual rainfall; high sunshine hours

B

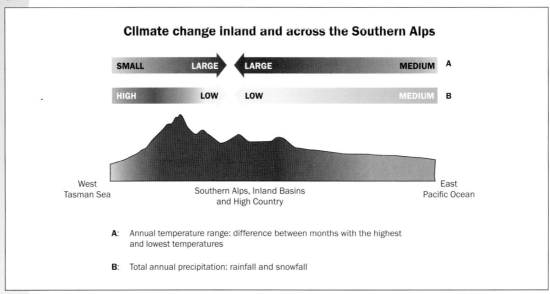

Climate change inland and across the Southern Alps

A: Annual temperature range: difference between months with the highest and lowest temperatures

B: Total annual precipitation: rainfall and snowfall

C

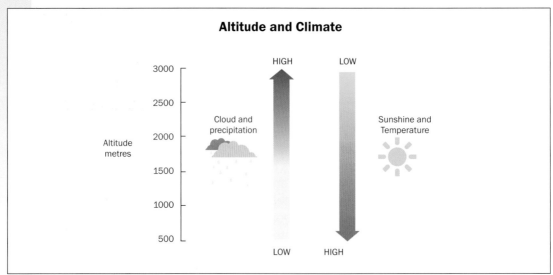

Altitude and Climate

Figure 44 Climate features of the Southern Alps and High Country — spatial variation due to location and altitude.

 ISBN: 9780170233316

A Location

Mount Cook Village and Lake Tekapo Village are just 45 km apart and at similar altitude. Mount Cook Village (the tourist village and visitor centre) is located east of the Main Divide but has features of the mountain climate shown in zone 2 of Figure 44A. By contrast, Lake Tekapo Village located at the southern end of Lake Tekapo has many features of the zone 4 continental-type climate shown in Figure 44A. The two locations have similar average monthly temperatures ranging between 15°C in summer and 3°C in winter, but the number of wet days, rainfall totals and hours of sunshine each year for the two places are very different.

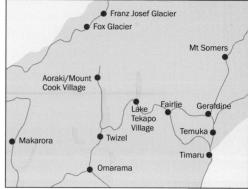

Figure 45 Comparing the climate of Mount Cook Village and Lake Tekapo Village.

B Annual data

	Mount Cook Village	Lake Tekapo Village
Number of wet days each year	164	79
Rainfall total each year (mm)	4491	591
Hours of sunshine each year	1570	2400

Figure 46

C Monthly climate statistics for Mount Cook Village and Lake Tekapo Village

	J	F	M	A	M	J	J	A	S	O	N	D
Rainfall Mount Cook Village (mm)	457	278	417	363	357	314	305	313	297	478	414	487
Rainfall Lake Tekapo Village (mm)	43	35	48	45	56	60	49	58	50	49	41	52
Temperatures Mount Cook Village (°C)	14.7	14.8	12.3	9.2	6.2	3.3	2.2	3.9	6.7	8.9	11.0	12.8
Temperatures Lake Tekapo Village (°C)	15.2	14.8	12.4	9.2	5.9	2.6	1.4	3.6	6.5	8.8	11.1	13.2

Figure 47

Learning Activities

1 In these six statements about the climate of the Southern Alps and High Country, there are three statements that are correct and three that are incorrect. Copy the three correct statements and write corrected versions of the three statements that are incorrect.

a North-facing hillsides get more sun and are warmer than south-facing hillsides.

b The west side of the Main Divide is much wetter than the eastern side of the Divide.

c The winds that blow most frequently (prevailing) in the Southern Alps and High Country come from the south and east.

d Inland basins like the Mackenzie Basin have very cold temperatures all year round.

e The nor'wester blows down from the mountains and brings a sudden spell of strong winds and warm and dry weather across the High Country.

f Where the sea and ocean has a big influence on the climate of the land it is called a 'microclimate effect'.

Learning Activities

2   Refer to Figure 44.

   a   Make a copy of Figure 44A and add these location labels to your diagram: West Coast; Main Divide; Southern Alps; inland basins; eastern High Country.

   b   Describe how the climate changes from west to east across the Southern Alps and High Country.

   c   Write a paragraph explaining why these changes take place. In your answer, you need to refer to different factors that influence climate.

3   a   Generalisation: Write one sentence to compare and contrast the climates of Mount Cook Village and Lake Tekapo Village.

   b   Draw graphs or diagrams to illustrate the annual statistics for wet days, rainfall and sunshine in Figure 46. Fully title your graphs/diagrams.

   c   Account for the annual cloudiness and precipitation (rain and snowfall) in these two places being so different.

Natural elements and natural processes in the Southern Alps and High Country environment — connections and interactions

Important understanding: Elements and natural processes interact within the Southern Alps and High Country environment. These interactions have helped to shape the environment. Movement and change happen all the time within the environment — this makes the environment 'dynamic'.

Elements

Elements are the parts of something. Elements of a natural environment are the landforms, geology, climate, soils and vegetation — these are the parts that together make up the environment. Elements are often referred to as the features of the environment. The elements themselves are made up of smaller sub-elements. Climate, for example, is made up of sub-elements like temperature, precipitation, sunshine, winds and humidity.

Processes

Processes involve a sequence of actions that shape and change the environment. In the Southern Alps and High Country environment, tectonic processes, weathering, glacial processes, fluvial processes and climate processes are examples of important natural processes that operate. Processes involve a number of elements. For example, the speed and amount of glacial erosion that takes place is determined by glacier depth and weight, the amount of material being carried in the sides and base of the glacier, and the strength of the rock surfaces the glacier travels over.

Interaction

Interaction means things being linked together and affecting each other. The landscape of the Southern Alps and High Country is the result of interaction between natural elements and natural processes. Interaction also causes the landscape to be continually changing. Processes operate all the time and as a result the landscape is always changing, both in the short term and long term. Process operation and landscape change are the reasons the environment can be called 'dynamic' — it is full of movement and change.

Examples of connections and interactions in the Southern Alps and High Country

- **A simple connection** is when one element has an impact on another element. For example, altitude affects vegetation, which results in vertical zonation of vegetation up a mountainside (Figure 38). With increasing height, temperatures decrease, but both rainfall and wind strength increase. Vegetation adapts to these changing conditions and at the highest altitudes where it is freezing all year round there is no or very little vegetation at all.
- **Combinations of different processes and elements** working together over time have shaped the landforms of the Southern Alps and High Country. For example, glacial erosion has resulted in the formation of U-shaped valleys all across the Southern Alps and High Country. The best-preserved examples are found today in the Fiordland area. In this area, the steep vertical sides as well as the flat valley floors have been preserved. Why? It is because the Fiordland area rocks are hard and tough igneous granite and diorite that have kept their steep vertical sides (Figure 48). In other parts of the Southern Alps and High Country, the rocks are weaker and more fractured. Here, frost action has weathered the valley sides and the vertical shape has been covered by scree and destroyed by landslides leaving steep, rather than vertical, slopes. Figure 20 highlights how the operation of internal and external processes has led to the formation of present-day landforms. U-shaped valleys provide an example of the way elements and processes operate together to shape the land.

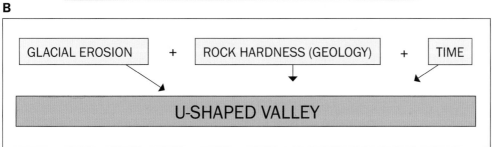

Figure 48 a and b U-shaped valley in Fiordland (Milford Sound) and U-shaped valley formation.

Interaction

i   **Relief affects precipitation**. The Southern Alps cause heavy falls of rain and snow by forming a barrier to the eastward movement of moist air from the Tasman Sea. The mountains force this air to rise in order for it to continue its eastward movement. As it rises it cools, rain clouds form and heavy rain and snowfall result on the western slopes of the Southern Alps and across the Main Divide. Some of this cloud, snow and rain spills over onto the eastern side of the mountains (orographic effect shown in Figure 35 on page 255).

ii **Precipitation affects relief**. This large amount of rainfall and snowfall results in the formation of powerful rivers and glaciers, which wear away and modify the landforms. The wide range of glacial landform features (described on pages 26–29) in the higher mountains and river-created landforms like outwash plains and V-shaped valleys (described on page 29) in the lower areas and intermontane basins are the result.

iii **Climate and relief interact in other ways**. Increasing altitude (relief) results in a lowering of temperatures (climate). The highest mountains remain freezing and snow-covered all year round. At lower altitudes, temperatures change from below freezing in winter to above freezing in summer. They also often drop below freezing at night but rise above freezing during the day. This constant cycle of temperature from above to below freezing results in weathering of rocks and hillsides by a freeze-thaw process. In turn, this leads to the formation of scree slopes and to frequent landslides.

Learning Activities

1 Simple connection: Draw a sketch of Figure 38. Annotate your sketch to highlight how altitude influences vegetation.

2 Copy Figure 48B but add extra information to each box so the diagram gives a brief explanation of U-shaped valley formation and the contribution each of the three elements makes.

3 Copy Figure 49. Make a key for boxes A and B with information about how relief affects precipitation (Box A) and how precipitation affects relief (Box B).

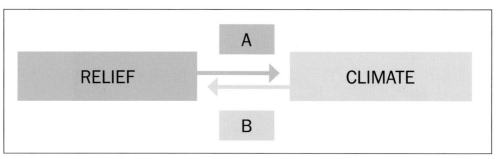

Figure 49 Interaction — relief and climate affect each other in the Southern Alps and High Country environment.

People in the Southern Alps and High Country

Important understanding:
- People perceive and use the Southern Alps and High Country environment in many different ways.
- Perceptions and uses have changed over time.
- There is a variety of interactions between people and the Southern Alps and High Country environment.

In this scarred country, this cold threshold land,
The mountains crouch like tigers. By the sea
Folk talk of them hid vaguely out of sight.
But here they stand in massed solidity
To seize upon the day and night horizon.

Men shut within a whelming bowl of hills
Grow strange, say little when they leave their high
Yet buried homesteads. Return there silently
When thunder of night-rivers fills the sky

— from 'The Mountains', *Jerusalem Sonnets: Poems for Colin Durning*, James K Baxter

 ISBN: 9780170233316

From a distance the Southern Alps and High Country environment appears to be completely natural — an environment with landforms dominated by mountains, glaciers, rivers, lakes and inland basins, with a land surface cover of forest, tussock grassland, bare slopes and rock. The harsh and cold climate of the area is reflected by the snow and ice that covers much of the ground surface in the highest areas all year round and during winter across the lower areas as well.

A closer view, however, shows evidence of people in the area. Roads, towns, tiny settlements with some intensive farming on lower land and in the basins can be seen when flying over the area. Zoom in closer and isolated farm buildings and livestock on large hill-country sheep stations can be seen, along with ski-lifts and chalets on snow-covered hillsides. Some of the lakes and rivers on closer inspection are revealed not to be natural at all but the work of people, with dams built and lakes formed to supply water to hydro-electric power stations, and huge canals built across the land to transfer water from one river to another.

Learning Activities

1 a How does James K Baxter describe landforms of the mountains in his poem? Use quotes from the poem in your answer.

  b Name a cultural feature referred to in the poem.

2 a List four features of the Southern Alps and High Country environment you see in a distant view and four features of the environment you see in a close-up view.

  b Create drawings to show these two sets of features.

  c How do a distant view and a close-up view of the Southern Alps and High Country create different impressions about the nature of the environment? This is a start to the answer: 'From a distance, the Southern Alps and High Country environment seems to .... Seeing and viewing the environment close up reveals different features ....,'

Interaction between people and the Southern Alps and High Country environment — people and the environment affect each other

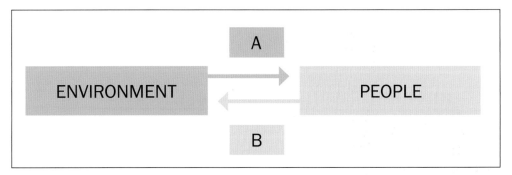

Figure 50 This diagram shows the two-way nature of interaction. It shows the environment affecting people (arrow A) and people affecting the environment (Arrow B).

Interaction example 1

In this first example, A and B are represented by two different aspects of environment — people interaction. The example for arrow A is how the nature of the environment influences the number of people who live in an area. The example for B is how people affect vegetation. The interaction relates to cultural and economic responses to the environment.

ISBN: 9780170233316

a Relief and climate (environment) have affected population distribution and population density in the Southern Alps and High Country area (Arrow A in Figure 50)

Few people live in the Southern Alps and High Country area — the mountainous terrain and inland location made this area difficult to access for both Maori and early European settlers. This same terrain plus the harsh climate have restricted farming, settlement and town growth. There are limited job opportunities in the area and no large towns to provide people with a full range of educational, health, social and cultural services and facilities. As a result of these challenging environment conditions, settlement has been limited and therefore population density is low throughout the area, at less than one person per square kilometre for most of the region. Across large areas of the land above 2000 metres and in the Fiordland, Mount Aspiring, Aoraki/Mount Cook and Arthur's Pass national parks, there is very little permanent settlement at all. The only large area with more than 10 people per square kilometre is the Queenstown Lakes district, which has a population of 30,000 people (Figure 51).

Towns and settlements are few in number and are located in lower areas, usually in valleys or beside lakes. Only Queenstown has a population of more than 10,000 people. There are a small number of towns with populations of between 1000 and 5000 people (Figure 51), but at most settlements such as Fox Glacier, Makarora, Lake Tekapo and Arthur's Pass, there are fewer than 500 permanent inhabitants. These are communities rather than towns.

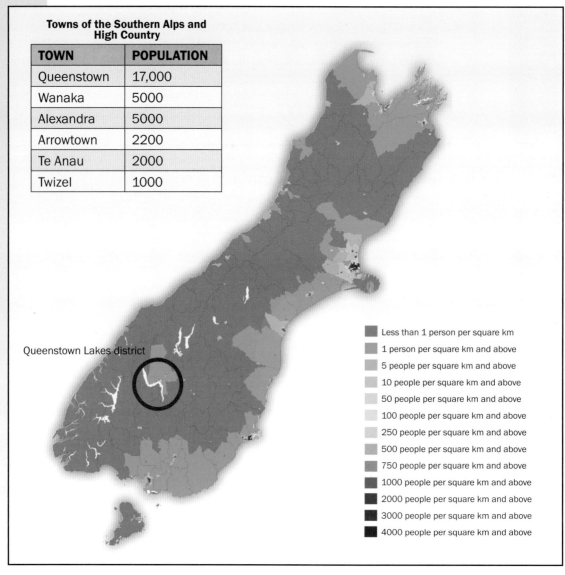

Towns of the Southern Alps and High Country

TOWN	POPULATION
Queenstown	17,000
Wanaka	5000
Alexandra	5000
Arrowtown	2200
Te Anau	2000
Twizel	1000

Queenstown Lakes district

Less than 1 person per square km
1 person per square km and above
5 people per square km and above
10 people per square km and above
50 people per square km and above
100 people per square km and above
250 people per square km and above
500 people per square km and above
750 people per square km and above
1000 people per square km and above
2000 people per square km and above
3000 people per square km and above
4000 people per square km and above

Figure 51 South Island population density.

 ISBN: 9780170233316

b People have altered the vegetation in the Southern Alps and High Country area (Arrow B in Figure 50)

Below the treeline both east and west of the Main Divide, beech forest was the natural climax vegetation. The beech forest was marginal further away from the Main Divide in the eastern High Country because of the low rainfall but it had still become established there. About 800 years ago people arrived. Maori lit fires to help with moa hunting and to clear routes for travel. These fires often got out of control in the dry conditions with strong winds blowing. Huge areas of forest were destroyed. In wetter areas west of the Main Divide, beech regenerated, but in the dry eastern High Country of inland Canterbury, the Mackenzie Basin and Central Otago, tussock species became more easily established and took control. Tussock replaced the beech forests.

European farmers seeing the eastern High Country for the first time in the mid-19th century thought the tussocks were the natural vegetation of the area and that these grasslands would be ideal for sheep. They set up huge sheep stations (very large farms) and continued to burn the tussock both to clear away sharp and unpalatable plants like speargrass and matagouri and to encourage new soft tussock shoots to grow. The regular burning plus heavy grazing by sheep resulted in the tall tussock species dying out with short tussock grasses becoming dominant. Introduced rabbits, which bred out of control and soon numbered many millions, became a major pest as they too ate the tussock. Many areas became so 'overgrazed' and 'overburnt' during the late 19th and first part of the 20th century that bare land with patches of low flat weeds like hieracium (Figure 52) and small wild pine trees became the new surface cover.

Figure 52 Hieracium (hawkweed).

A government report summed up some of the changes to the vegetation: 'Tussock has many of the characteristics of a forest … it is the product of a long slow development, and like a forest it is much easier to destroy than rebuild.'

Since the 1950s, vegetation change has continued. Farmers have adopted new land-management systems and have not grazed the land so heavily. Grazing has stopped altogether on much of the higher land. Rabbit control has been more successful. On some of the lower land and flatter land, topdressing, oversowing with grasses and clover, and irrigation has taken place. The result is that there are now three main types of land surface cover below the tree line in the High Country (Figure 54):

i   regenerating native tussock grasslands
ii a lot of higher and steeper land that is still very degraded with bare ground and flat weeds dominant
iii areas of improved and planted pastures on lower, more intensively farmed land (Figure 53).

Figure 53 Sheep grazing on improved pasture in lower and flatter High Country.

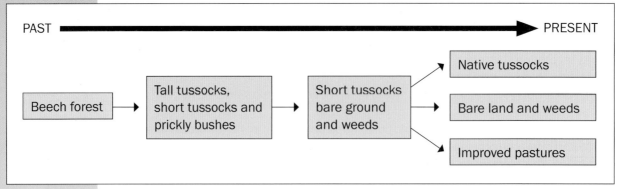

PAST ➤ PRESENT

Beech forest → Tall tussocks, short tussocks and prickly bushes → Short tussocks bare ground and weeds → Native tussocks / Bare land and weeds / Improved pastures

Figure 54 Vegetation — an overview of change.

Learning Activity

1   Using population distribution/density and the influence of people on vegetation in the Southern Alps and High Country as the examples, explain fully how people and the environment interact. Quote specific Southern Alps and High Country evidence to support your answer.

Interaction example 2

This example is about the way people have responded to the natural environment and the way this response has then influenced the way people use the environment. National Parks and World Heritage areas are the focus of this case study. People have placed high value on the beauty of the natural environment of the Southern Alps – High Country area. This has resulted in the government establishing national parks in the area. Development within the parks is restricted, so that the natural environment is protected and preserved. Access into and through the parks by people is controlled. Track building and hut construction to accommodate trampers is carried out in a way that is friendly to the environment. Environment protection and conservation take priority over economic development in the parks. Fiordland is New Zealand's largest national park and one of the largest in the world. Fiordland along with Aoraki/Mount Cook National Park and Mount Aspiring National Park form an area in the southwest of New Zealand called Te Wahipounamu. This area gained international recognition as a World Heritage area for its outstanding natural features. Remnants of the ice age, links with ancient Gondwana, ongoing geological processes, the rugged coastline and spectacular fiords, areas of lowland rainforests and wetlands providing habitats for rare plant and animal species are the reasons why South West New Zealand (Te Wahipounamu) qualified for the award of 'World Heritage Area'.

Learning Activities

1   What and where is Te Wahipounamu?

2   What environment features led to Te Wahipounamu being awarded World Heritage status?

3   How does setting up national parks or being awarded World Heritage status influence the way people use and manage places?

4   Draw sketches of Figures 56 and 57. Choose and copy an appropriate sentence from the text on page 50 to describe what each photo is showing.

Figure 55 Fiordland: part of Te Wahipounamu World Heritage area.

Figure 56 Across Fiordland National Park on the Milford Track.

Figure 57 One of many waterfalls in Fiordland National Park.

Interaction example 3

This last example is about people, dams, lakes and canals, with a focus on hydro-electric power (HEP) developments. Rising demand for power across New Zealand during the 20th century meant people were looking for ways to increase the supply of electricity. They saw the environment of the Southern Alps and High Country as having the potential for generating huge amounts of power from the water resources of the area. The run-off from the high rainfall and snowmelt into rivers and lakes, as well as deep and long valleys, were seen as providing ideal conditions for the generation of cheap HEP. People responded to these conditions by building dams and power stations in the Southern Alps and High Country area. People's actions and government decisions to go ahead with the HEP schemes were influenced by both environmental and economic factors.

Manapouri in Fiordland became the location of New Zealand's largest single power station, and the Waitaki River hydro scheme with its series of interconnected lakes, canals, dams and power stations now produces more electricity than any other New Zealand river. The landscape has been completely altered in the places where these power projects have taken place — people have had a huge impact on the environment.

ISBN: 9780170233316

Case Study 2

The Mackenzie Basin – Upper Waitaki HEP developments

The Mackenzie Basin (Figure 58–61), located at the centre of the upper Waitaki River power scheme, has been transformed by the HEP developments. These developments have resulted in a river and lake system that is one of the most modified in the country. It is a landscape that highlights how people have altered landforms and put their mark on the landscape. The ribbon lakes Tekapo, Pukaki and Ohau occupy U-shaped valleys gouged out by glaciers and dammed by terminal moraines. These lakes naturally feed water into the upper Waitaki River system. But there nature ends and people have taken over. Dams have been built which have raised lake levels, created new lakes like Lake Ruataniwha, and canals constructed (Figure 58, 59 and 60) to redirect water to provide a large and constant supply of water to the HEP stations. The construction of the 50 km of water canals across the Mackenzie Basin involved the largest earthmoving project ever carried out in New Zealand, and the huge Benmore Dam resulted in the formation of New Zealand's largest human-made lake, Lake Benmore (Figure 60).

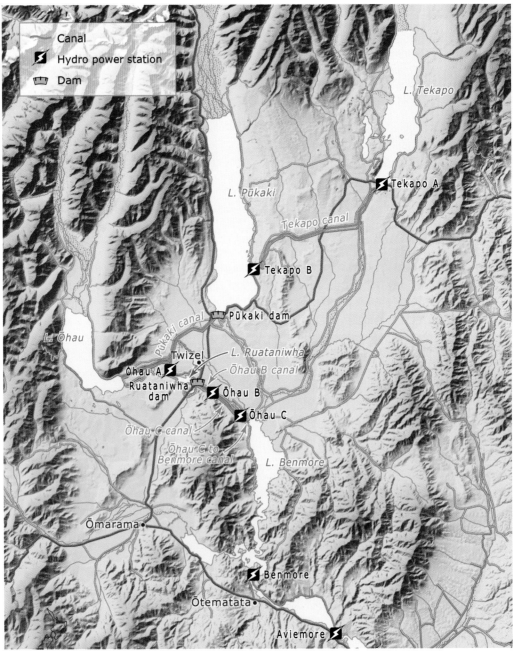

Figure 58 The Mackenzie Basin and HEP developments.

Case Study 2

Figure 59 The Tekapo canal transfers water from Lake Tekapo to Lake Pukaki.

Figure 60 HEP; looking back towards the Southern Alps. In the foreground are the Waitaki River, Benmore power station, dam and artificial Lake Benmore.

A	Lake Pukaki	**B**	Pukaki Canal
C	Lake Ohau	**D**	Lake Ruataniwha
E	Pivot irrigation	**F**	Ohau Canal and canal-side irrigation
G	Lake Benmore	**H**	Mackenzie Basin

Figure 61 Twizel and the surrounding Mackenzie Basin.

1 **a** The environment influenced the decisions of people: What were the natural resources that led to people developing huge HEP schemes in the Mackenzie Basin? Write a short paragraph answer.

b HEP developments affected the environment: What influence did the HEP development have on the environment? Write a detailed answer and include one visual (annotated sketch, map or diagram) to support your answer.

Twizel and Queenstown — two contrasting towns

Twizel was a town built as a home for workers employed on the Upper Waitaki HEP scheme. When the scheme was begun in 1968 there was no Twizel. By 1976 a 'new town' had been built and had a population of 6000 — mostly workers and families connected with the HEP scheme, plus people providing services like shop and business owners, bank staff, health workers and teachers. Twizel was a self-contained town, providing all the services and facilities the population needed. However, the population of Twizel went into rapid decline once the HEP scheme was completed. By 1990, four years after the scheme was finished, the population had dropped back to 1000 people. It appeared the town might cease to exist.

Local people fought to keep the town alive. A few people were needed to operate and maintain the hydro scheme and many of these wanted to live in Twizel to be close to their work. Some former scheme workers wanted to retire in Twizel. The town had spare houses and infrastructure like roads, parks, water and electricity supply in place. Cheap houses attracted people from places like Christchurch to buy holiday homes in Twizel, while others came to settle permanently and start businesses. These businesses made use of the natural and cultural resources of the Mackenzie Basin for outdoor recreation, including rowing on Lake Ruataniwha, fish farming and more intensive land use on irrigated land including dairy farming (Figure 61).

Since 1990, the population has remained around a thousand. Over Christmas and New Year, people coming on vacation means the town size grows to as many as 5000 for a few weeks of the year. For the rest of the year, the town struggles on. It has not been able to transform into a major tourist town or base for winter skiing as had been hoped. Neither has it developed as a farm service town. Many of the services the town once provided like a full range of medical and educational facilities and shops have closed down. Future growth and prosperity remain uncertain.

Figure 62 Tourism in the Twizel area.

Queenstown has a much longer history than Twizel. By the time the building of Twizel started in 1968, Queenstown had already celebrated its first 100 years. Gold discoveries had led to the foundation of the town, but for much of the 20th century the population of Queenstown was little more than 1000 people. In the 1970s, more people were living in Twizel than in Queenstown. Even in 1981 the population of Queenstown was less than 3500. Since then the fortunes of the two towns have been very different. Queenstown

Case Study 3

has boomed and now has a population of close to 20,000. The boom has been based on tourism. Over one million tourists stay in Queenstown each year and another million come as day visitors. Queenstown's location on the shore of Lake Wakatipu surrounded by mountains gives it instant appeal — it has the 'wow' factor. The mountains around the town have been the sites of skifield developments like at Coronet Peak. Adventure tourism businesses have been set up using the Shotover and Kawarau Rivers with their deep gorges for white-water rafting, jet boating and bungy jumping. Four-wheel-drive vehicles and mountain-bike trails give access into the historic surrounding area and remains of the old goldfield towns and workings. Bus tour groups and independent backpackers are both catered for. Queenstown and the surrounding area have developed into a place where tourism is a four-seasons business. There are cultural, arts and sports events and festivals in every month of the year. Sealed road access, airport development and tapping into the growing domestic and international tourism market have underpinned Queenstown's growth. The town and surrounding region have become one of the fastest growing areas in the country. Further airport expansion is planned, and new hotels, new shopping areas, new schools and new medical facilities are all being built. The future of the town seems assured.

Figure 63 a–d Queenstown adventure tourism.

Learning Activity

1 Write a script for a TV news report that contrasts the fortunes of Queenstown and Twizel. Include these four things:
- a brief overview of the population growth of the two towns
- explanation of the different fortunes of the two towns
- a look into the future with a prediction of what the towns might be like in 2050
- visuals and images you would use in the programme.

Perception and use of the environment — how people view the environment and how these views have changed over time

Perception is about how people see things and the views and opinions people have about things. In Geography, perception is to do with the way people view and interpret places and the environment and how they think about and view geographic issues.

Perception is often influenced by **perspective** — who we are and the position we come from influences the way we view places, the environment and the opinions we have about geographic issues.

The interaction between people and the natural environment illustrated by HEP projects in the High Country is also an example of how the way people perceive the environment does not always remain the same. People only came to see and value the Southern Alps and High Country for its HEP potential during the 20th century when demand for electricity grew and technology became available to make use of the water resources of the area to generate and distribute electricity. In the 21st century, concern about damage to the environment caused by huge HEP projects has grown. This means proposals for new HEP projects and other developments in the Southern Alps and High Country meet opposition from people concerned about sustainability, upsetting the balance of nature and wanting to conserve high-value wilderness and scenic areas of this environment. There is tension between the demands of economic growth, social well-being and environmental protection.

From being seen as a harsh, remote and stunning environment during the 19th century suited to sheep grazing (Figure 64) and sightseeing with some hunting, fishing and climbing, perception of the Southern Alps and High Country has changed. In the last hundred years, and especially since the 1970s, people have perceived the environment of the Southern Alps and High Country as offering opportunities for development and use in an ever-increasing number of ways (Figure 78). At the same time, people have also placed increasing emphasis on preserving the natural beauty and resources of the area, and on using the environment in sustainable ways.

Figure 64 Sheep grazing in the High Country.

The tension described above is plain to see and hear with different people having different priorities.
- In the 19th century, farmers saw the tussock hill country as land to use and from which to make a profit. The environment was often exploited without any recognition of damage being done to the vegetation and soil. This farming system was not sustainable. Farmers now see themselves as inheritors and guardians of the environment, while at the same time having a livelihood to make and need to run their farms in a profitable but sustainable way. They have experience and expertise and feel they know best how to use and manage their land.

- The government recognises the importance of having economically successful farms that provide valuable wool, sheep meat and other farm produce exports. They want to support ways of maintaining and increasing farm production, for example through the increased use of irrigation, and to encourage tourism growth. At the same time they want to conserve the environment and protect the natural heritage of the area. The government tries to strike a balance between economics and the environment.
- Tenure review began during the 1990s. It involves farmers with large high-country sheep runs on land leased from the Crown (government) negotiating with the Crown to get freehold ownership of some of their farm, usually the better lower land. In return, they give up their lease of land of high visual, historic, ecological, scientific and cultural value (usually the higher land). This land then becomes owned by the Crown, is retired from farming and becomes conservation land. The public get access to this land. Tenure review has the aim of allowing for the full economic potential of the better farmland to be realised, while also conserving and giving public access across the High Country. Satisfying the priorities and needs of farmers, the government, recreation groups and conservationists is not always easy.

Figure 65 Helicopter tourist access.

- The public want to have easy access to the mountains, the rivers and the snow. They want not only to be able to see the area but also to use it for recreational activities. They regard the area as a national treasure that everyone owns and has a right of access to (Figures 65 and 66).
- Conservation and environment groups, while recognising the rights of farmers and tourists to use the Southern Alps and High Country, place environmental protection as a priority. They value the unique landforms and the flora and fauna of the environment. They view the national parks and conservation lands of the area as serving a vital purpose and view development proposals within the environment in a critical way.

Figure 66 Hunting for tahr in the Southern Alps.

Summary of changes in the way the environment has been used

1 More diversified types of farming and more intensive farming of some land taking place alongside the traditional extensive sheep farming.
2 Increasing tourism across the region to the point where tourism and farming are equal in terms of employment and income generated.
3 HEP schemes have had huge impacts on the natural environment and have also brought about social and economic change.
4 Growth in conservation and a focus on sustainable ways of using the environment. A lot of higher land has been retired from farming and conservation areas have been set up.

ISBN: 9780170233316

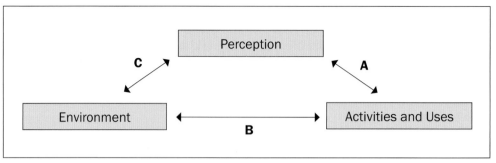

Figure 67 Interaction between the way people perceive the environment and the way the environment is used.

Why perception has changed

Economics

Falling and fluctuating prices for wool and sheep meat has meant High Country sheep stations have become less profitable. Farmers have looked for new ways of gaining income and profit from the land and resources of the area. Irrigation and more intensive farming of lower farm areas and developing on-farm tourism are two ways farming has diversified. The potential of the environment to attract tourists and be the basis for tourist businesses to grow has been recognised.

Research and knowledge

Scientific research and market research identified opportunities to farm beef cattle and deer, which can be more profitable than sheep. Viticulture is a new industry for this region. Central Otago became the world's most southerly wine-producing region, with successful wine-grapes planted in large quantities only since the late 1970s. The potential for viticulture had been known for over a hundred years but successful small plantings in the 1970s led to commercial-scale grape planting, especially of pinot noir grapes, and the opening of many small wineries. Small-quantity but high-quality wine production has been the focus.

Access

Dead-end clay tracks and gravel roads in and out of the region made access difficult for a long time. Development of a sealed-road network has taken place only since 1950 (Figure 68). Only three roads cross the Southern Alps: Arthur's Pass, Haast Pass and the Lewis Pass. This description of State Highway 73, the Arthur's Pass highway, highlights road improvements that have taken place:

In earlier times, the teams of horses on Cobb & Co coaches struggled up the steepest slopes while the passengers walked, and the brakes were jammed on for the descent. Then came the years when buses had to back to manoeuvre round the sharp bends. Now, the 13-metre vehicle length limit is gone, the worst hazards are removed and except when winter conditions close the route, modern cars, buses and trucks travel easily over sealed roads.

Air and helicopter flights with commercial operators like Air New Zealand (to and from Queenstown) and smaller private companies operating out of airfields like those at Tekapo and Aoraki/Mount Cook have improved access and reduced travel times into the area as well as offering scenic flights to tourists. Now, the Internet, computers and smart phones have opened up the area to the world and the world to the area. Farmers can gain instant access to suppliers and to markets; tourists can research attractions and book trips and tours online. E-access means the remoteness and distance are no longer the barriers they once were.

Figure 68 Crossing the Waimakariri River – Highway to Arthur's Pass.

Environmental protection and sustainability

As people globally have become more concerned about the environment, this mountainous region with its spectacular landscapes, sense of remoteness and unique and fragile plant and animal life has been viewed through a different lens. People place great value on protecting the environment, while at the same time wanting the public to have access to it. Big development schemes that are money and profit focused, such as expanding intensive dairy farming with cows housed under cover, are viewed unfavourably by people who have a conservation focus and love for the natural environment. At the same time, these and other people argue the public has the right of access into and across the area for recreation.

Learning Activities

1 **Changing perception**

   a   How did people perceive and value the Souther Alps and High Country environment in the mid 1800s? How do people perceive and value the environment today?

   b   Four reasons are given on pages 58–59 explaining why peoples' perception of the Southern Alps and High Country environment has changed over time. List these four reasons or show them in a star diagram.

   c   Which of these four reasons do you think has been the most important one in bringing about the perception changes? Justify your answer.

2 **Different viewpoints**: Why would farmers and tourists view and value the Southern Alps and High Country environment in different ways?

3 **Important idea**: 'There is tension between the demands of economic growth, social well-being and environmental protection.' This is a statement from page 56. Expand upon and explain this statement by using specific examples from the Southern Alps and High Country environment.

4 **Perception and use of the environment**: Make a copy of Figure 67. For each arrow letter (A, B and C), give two examples from the Southern Alps and High Country environment and events that would match and illustrate that arrow. Examples:

   •   'Improved access with sealed highways like across Arthur's Pass and Internet use means people view the area as accessible and less remote. More people visit the area as tourists' would fit with Arrow A.

   •   'Winter snow on higher hillsides has resulted in skifield developments, for example at Coronet Peak' would fit with Arrow B.

Glentanner Station – perception and change over time

Location

Glentanner is located deep in the High Country on the west side of Lake Pukaki on the road to Mount Cook Village. The location provides good access to the Tasman River and to the lakes, mountains and glaciers of Aoraki/Mount Cook National Park. Mount Cook Village is a 15-minute drive away. Much of the land is rough mountainous country above 1000 metres altitude, but has the advantage of facing north and so gets more sun, and snow does not lie as long as on south-facing slopes. The lower land of the station is in the river valley and close to the lake shore.

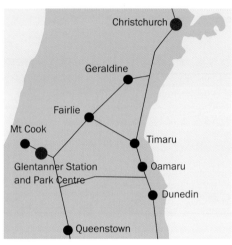

Figure 69 Location of Glentanner Station and Park Centre.

1858

The first runholders, the Dark brothers, brought merino sheep onto the land at Glentanner. Just getting sheep onto the farm was difficult. The flocks were driven up from the Canterbury lowlands with few tracks to follow, through tangled tussock and sharp matagouri scrub with dangerous rivers to cross. Early times were difficult because of uncertain prices for wool, stock losses during heavy winter snows and isolation. Rabbit infestations added to the problems of finding enough pasture for the sheep to eat. Bullock wagons were used to cart in supplies and carry out wool. The farm changed owners several times.

Figure 70 On the road to Glentanner and looking towards Aoraki/Mount Cook from the west side of Lake Pukaki.

 ISBN: 9780170233316

Case Study 4

1957

The present owners, the Ivey family, took over the farm. The farm covers an area of 18,000 hectares and produces fine wool from 9000 merino sheep. The farm also runs deer and cattle. The farming operation provided the income from the farm. Farm management has been to make sure the farm operates both profitably and sustainably. Some lower land on the flats has been irrigated and pastures improved. This has allowed stock to be removed from higher scenic areas to help preserve the natural environment there. The spread of unwanted pine trees onto the tussock lands has been a continual battle for the farmers.

1970

The Glentanner Park Centre was opened. This tourist centre is set within the working Glentanner sheep station.

Figure 71 Glentanner Station and Park Centre — marketing of the farm and tourist centre.

Diversification

Glentanner has become famous for expanding into tourism. This expansion was demand and economic driven. The demand has come from ever-increasing numbers of tourists visiting the Aoraki/Mount Cook area, estimated to be 350,000 each year with over 3000 every day over summer. Although many are day trippers, others want to experience high-country farm life and the high-country environment close up, and have a range of activities available from passive, such as sightseeing and photography, to active and adventurous, such as skiing and mountain climbing. Diversifying into tourism has made Glentanner more economic by providing a wider range of income sources which make best use of the natural resources of the farm and surrounding area. Glentanner has taken advantage of having a lakeside location on the main road leading up to Mount Cook and the Hermitage and providing easy access into the Aoraki/Mount Cook National Park.

Figure 72 The traditional — sheep on the road to fresh pasture between Glentanner and Mount Cook.

Figure 73 a and b Reception centre at Glentanner Park.

- The Iveys are passionate environmentalists who take great care of their land and animals. They feel lucky to be living in such a beautiful part of the world. Thus, sustainability and best farming practices are very important to them.

- Affordable accommodation including self-contained motel style units, standard units, basic units, backpackers, powered sites and camping sites — with breathtaking views of Aoraki/Mount Cook, Mount Cook National Park, Lake Pukaki and Glentanner Station.

Figure 74 Mountain biking across Glentanner Station.

- Awesome outdoor adventure activities for everyone. Adventure activities include flight-seeing, helicopter flights with snow landings, horse trekking, glacier trips, 4WD tours, heli-skiing, mountain biking, fishing, hunting, ice climbing, hikes, walks and farm tours. Glentanner is also located on the Alps 2 Ocean Cycle Trail.

Figure 75 Heli-skiing: dropped by helicopter to ski high above Glentanner.

Figure 76 Skiing above Glentanner Station.

 ISBN: 9780170233316

Case Study 4

Figure 77 Horse trekking at Glentanner.

Learning Activities

1 Using Figure 69 and an atlas or online e-map, draw your own sketch map to show the location of Glentanner Station and Park Centre.

2 How has the use of Glentanner changed over time? Start by describing the early years of the farm, 1850–1900, and finish by referring to the present-day Park Centre developments.

3 **a** Complete this word mix by fitting the words from the box across the name of GLENTANNER.

```
                    G
                    L
                    E
                    N
                    T
         P U K A K I
                    N
                    N
 E N V I R O N M E N T
                    R
```

| IRRIGATION |
| WOOL |
| DIVERSIFICATION |
| HELI-SKIING |
| TUSSOCK |
| PUKAKI |
| MERINO |
| SUSTAINABILITY |
| ENVIRONMENT |
| TOURISM |

b Write a sentence for each word from the box about Glentanner. For example, *'Glentanner is located on the western side of Lake PUKAKI.'*

Learning Activity

1 **Summing up:** Refer to the information in Figure 78. Write paragraphs describing and explaining how the way people perceive and use the Southern Alps and High Country has changed over time.

People's use of the Southern Alps and High Country — increasing variety over time

PAST 1200AD–1850	19th CENTURY 1851–1900	20th CENTURY 1901–1950	20th CENTURY 1951–2000	21st CENTURY TO PRESENT 2001–2013
Maori: food gathering and hunting for eels, weka and moa; obtaining pounamu; east coast to west coast travel routes				
European awareness and exploration				
	Gold rushes — Central Otago			
	Farming: **large sheep farms** (sheep stations) set up and operate across the High Country; merino sheep			
			Beef cattle and deer introduced on high-country farms	
			Irrigation of lower and flatter land in basins and valleys; sown pastures **more intensive farming** + some dairying in the Mackenzie Basin; viticulture and wine production in Central Otago	
			HEP projects: Upper Waitaki, Clutha and Manapouri dams and power plants.	
		Tourism: hunting, fishing, tramping, climbing, mountaineering		

Figure 78 People and the Southern Alps and High Country.

 ISBN: 9780170233316

Tourism: skiing and ski-field developments – Queenstown, Wanaka, Coronet Peak, Remarkables, Cardrona, Mount Hutt

Snowboarding and adventure tourism, e.g. heli-skiing, paragliding, river rafting, jet boating, bungy jumping

Ecotourism, e.g working farm tours and farmstays, guided nature walks and tramps, a focus on native plant and animal life, landforms and history

Coach tours, air and helicopter sightseeing

Multisport events: running, kayaking, cycling; mountain biking; endurance events

Rapid **town growth**, lifestyle block subdivision and holiday home building, e.g. Queenstown, Wanaka, Lake Tekapo

National Parks: Arthur's Pass established

Mt Aspiring NP, Fiordland NP, Aoraki/Mt Cook NP established

Te Wahipounamu declared a World Heritage area

Tenure review: some sheep stations divided into freehold land for farming, with higher land and high-value landscape land retired from farming and becoming conservation land

Off-road cycle trails and cycle touring; wine tours; film set tours

Astronomy: Mt John Observatory

Aoraki-Mackenzie International Dark Sky Reserve

Mackenzie Basin: agreement between community groups, landowners, farmers, conservation groups, environment experts and irrigation groups to work together to manage conservation, tourism and farming in the basin in a sustainable way

Spinifex plants in the Red Centre: the Great Victoria desert.

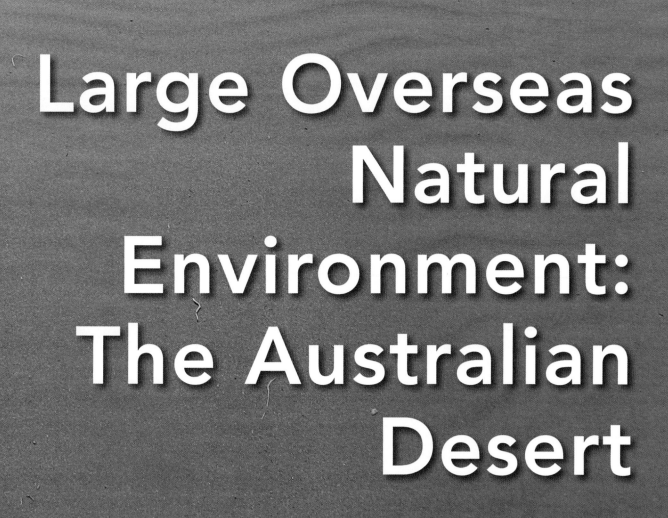

Large Overseas Natural Environment: The Australian Desert

2 A large overseas natural environment: The Australian desert

Iconic feature of the Australian desert: Uluru

Important understanding: Uluru, also known as Uluru/Ayers Rock, is one of Australia's most famous and treasured landform features. It is a located in the centre of Australia in the heart of the Australian desert environment.

Uluru is part of the Uluru-Kata Tjuta National Park in the southeast corner of the Great Sandy Desert (Figure 1 and 2). It is an iconic feature of the desert environment area and has World Heritage status for its cultural and natural importance and distinctive appearance.

Figure 1 Uluru/Ayers Rock.

 ISBN: 9780170233316

Figure 2 Uluru from above. The rock has been scarred into deep grooves and caves by water and wind erosion over millions of years. Trees and other vegetation around the base of the rock make it look like there are streams of turquoise water flowing from the base of the rock. There is little vegetation on the rock itself.

Ten facts about Uluru

- Uluru is not that large, at just 348 metres high and 9.5 km around. Its greatest diameter length is 3.6 km and it covers an area of about 3.3 sq km. The climb to the top is 1.6 km long, much of which is at a steep angle. The flat summit and steep sides are full of distinctive surface features like valleys, grooves, ridges and caves that are the result of millions of years of erosion.
- It is a block of red-coloured sandstone. This colouring, together with being located in a flat plain, makes Uluru stand out and be visible from far away, both at ground level and from the air. The red surface colouring is caused by oxidation of the iron content of the sandstone when exposed to air. Without the oxidation, the rock would be grey coloured.
- Uluru is not really a hill or a mountain. It is sometimes called a monolith — a huge single piece of rock. It is one of the largest monoliths in the world.
- Geographers and geologists describe Uluru as an inselberg, which means an 'island mountain' — a steep-sided hill made out of hard rock. An inselberg is a landform that is the final remains of a mountain range that has mostly been eroded away. Uluru is very old — its formation began more than 500 million years ago.
- The Uluru area has desert climate features: low annual rainfall of less than 250 mm, and temperatures that can be a burning 47°C during the day in summer, with night temperatures in winter dropping below freezing to as low as −8°C.
- The local Aboriginal people, the Anangu, divide the year into five weather seasons:
  - a Wanitjunkupai (April/May) — cooler weather
  - b Wari (June/July) — cold season bringing morning frosts
  - c Piriyakutu (August/September/October) — animals breed and food plants flower
  - d Mai Wiyaringkupai (November/December) — the hot season when food becomes scarce
  - e Itjanu (January/February/March) — sporadic storms can roll in suddenly.
- Uluru was named Ayers Rock in 1873 by European explorers, after the Premier of South Australia, Sir Henry Ayers. The rock resumed its original name in 1985 when the land was returned to its traditional owners, the Anangu. Although the name of Ayers Rock is still used, as is the dual name of Uluru/Ayers Rock, the official name of the rock is once again Uluru.

- The nearest large town to Uluru is Alice Springs, which is more than 300 km away to the northeast. The road journey from Alice Springs to Uluru takes around five hours. Travel by air takes one hour for the same route.
- Tourists from all over Australia and the world come to see the rock and watch spectacular sunrises and sunsets at the rock site. Around 350,000 tourists visit each year. As many as 100,000 of these climb Uluru, although this is disapproved of by many Anangu and other Aboriginal people for whom Uluru is a sacred site.
- Aborigines first moved into and settled in the Uluru area about 10,000 years ago. The Aboriginal understanding of Uluru is as follows: 'In the beginning the world was unformed and featureless. None of the places we know existed until ancestors, in the form of people, plants and animals, travelled across the land. As they travelled they formed the world as we know it by creating all living species, and the rocks, caves, boulders, cracks and waterholes of the desert landscape that are present today. Uluru is physical evidence of acts carried out during this creation period. Anangu see themselves as the direct descendants of these ancestral beings and are responsible for the protection and appropriate management of these ancestral lands.' The knowledge and stories about the land and people have been passed on down the generations in stories, songs, dances and ceremonies.

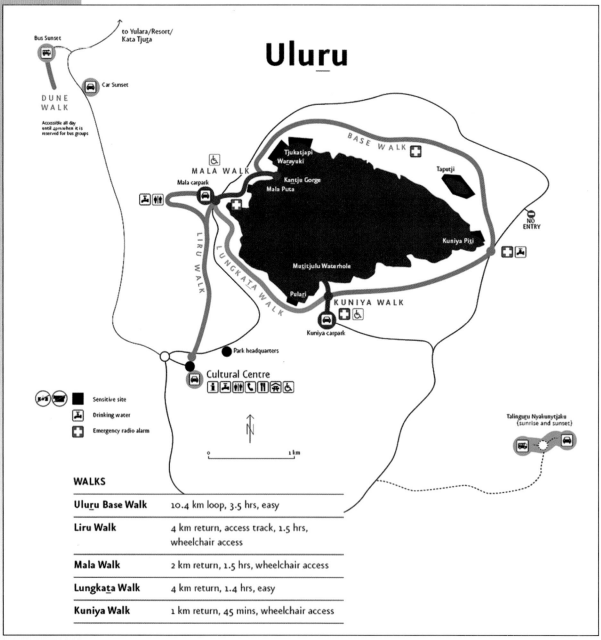

Figure 3 Uluru — a tourist attraction.

WALKS

Uluru Base Walk	10.4 km loop, 3.5 hrs, easy
Liru Walk	4 km return, access track, 1.5 hrs, wheelchair access
Mala Walk	2 km return, 1.5 hrs, wheelchair access
Lungkata Walk	4 km return, 1.4 hrs, easy
Kuniya Walk	1 km return, 45 mins, wheelchair access

Learning Activities

1 **a** In which Australian desert is Uluru located?

  b From above, what looks turquoise coloured around the base of Uluru?

  c How far away from Alice Springs is Uluru? Give an answer in kilometre distance and travel time.

  d Uluru is located in the land area of which Aboriginal people?

  e What is 'easy', 10.4 km long and takes 3.5 hours?

2 What natural and cultural features of Uluru do you think would have led to it being given World Heritage status?

3 Either write a 30–40-word description of Uluru or draw an annotated sketch of it.

4 Why is Uluru valued and treasured so much by the Anangu?

How the rock got there and why it looks the way it does

Uluru is the exposed remains of a sandstone sedimentary rock called arkose. The rock was formed out of sand eroded from nearby mountains that were first deposited in a large and deep fan at the base of the mountains. About 500 million years ago this fan was covered by rising sea and had deposits of mud build up on top of it. The sand was compressed by the sea and mud into a hard sandstone rock. Since then the hard sandstone has been folded and tilted upwards. Weaker rock around it has then been eroded away by weathering and erosion, leaving Uluru alone protruding out of the desert.

Uluru and Kata Tjuta (also called the Olgas) are twin features of the Uluru-Kata Tjuta National Park. Kata Tjuta (Figure 4), like Uluru, is a famous landform feature and tourist attraction. Also like Uluru, it is an inselberg. These two inselbergs are closely related and began their formation 600 million years ago. They are thought to have originally been one massive monolith rather than the 36 separate domes they are today.

Figure 4a and b Kata Tjuta.

Uluru: development over time

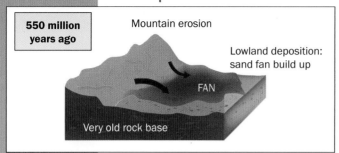

Figure 5

- 600–550 million years ago there were high mountain ranges to the west of Kata Tjuta and Uluru. Rainwater flowing down these mountains eroded sand and rock and dropped it in big fan shapes on the surrounding plain. One fan was made up mostly of stones, gravels and pebbles. The other fan was mainly sand. Both fans over time grew to become kilometres thick.

- 500 millions years ago the whole area became covered by the sea. Mud fell to the bottom of the sea and covered the seabed, including the fans. The weight of the new seabed turned both the mud and the fans beneath it into rock. The stone, gravel and pebble fan became conglomerate rock — this was the future Kata Tjuta. The sand fan turned into a coarse sandstone rock called arkose — this was the future Uluru.

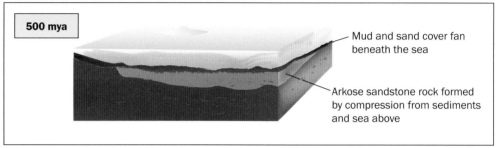

Figure 6

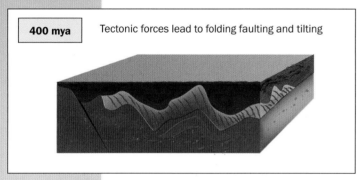

Figure 7

- 400 million years ago, the sea had disappeared and huge tectonic earth forces pressed in on Central Australia. Some rocks folded and tilted. The conglomerate fan (Kata Tjuta) tilted just a small amount, about 20 degrees. The sandstone fan (Uluru) tilted much more, as much as 90 degrees, so the layers of sandstone almost stood on end.

- Over the last 300 million years, the mudstone and other softer rocks that covered and surrounded the conglomerate and sandstone blocks have been eroded away by water and wind. The old hard conglomerate and sandstone rocks have resisted erosion and have been left standing above the surrounding eroded land. These resistant blocks of rock are Kata Tjuta and Uluru.

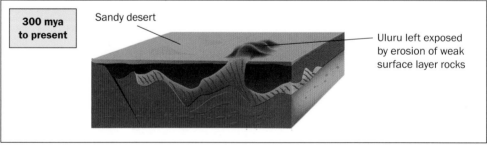

Figure 8

- 30,000 years ago the area around Uluru and Kata Tjuta became covered in windblown sand plains and sand dunes, so the two inselbergs are now set in an area of mostly flat sandy desert.

The shaping of the surface of Uluru and Kata Tjuta

The photos in Figures 1, 2, 4 and 9 show that both Uluru and Kata Tjuta, although appearing to be smooth and featureless from a distance, are both, in fact, full of caves, holes, grooves and ridges. Anangu see these features as markings caused by journeys and actions of ancestral beings across the landscape. Kata Tjuta, the Anangu name for the collection of domes, means 'many heads'. Geologists see these features as evidence of weathering and erosion eating into the surface of the inselbergs during times of wetter past climates and continuing today.

Kata Tjuta is made up of distinctive rock domes. Tectonic earth movements in the past caused vertical fractures (cracks) through the conglomerate rock. Water seeped into the cracks and over millions of years the rock slowly weathered and eroded away to form valleys and gorges that split the rock slab into blocks. These blocks were then further weathered and eroded away to produce the rounded domes that are present today.

Uluru has parallel ridges and grooves running across its surface. These are the result of softer layers of the sandstone being worn away more quickly than other harder bands in the rock. Rainstorm after rainstorm over millions of years has sent water across and down the sides of the rock, cutting out grooves and chains of potholes and plunge pools. Caves at higher levels on Uluru give the rock a honeycomb look; at ground level the caves have a wave shape (Figure 9). The caves are the result of chemicals in rainwater eating into the sandstone rock.

The surface of Uluru is covered with a lot of flaky pieces of rock. These are caused by chemical weathering and mechanical weathering. Chemical weathering results from water and oxygen in the air causing minerals in the rocks to decay. Mechanical weathering happens when hot sun causes rocks to expand during the day and rapid cooling at night leads to rocks contracting. This constant expansion and contraction causes pressure change, which stresses the rocks and over time causes them to crack, splinter and break up.

Figure 9 Caves, valleys and grooves on the side of Uluru.

ISBN: 9780170233316

1   Uluru began life as a part of a large mountain range 600 million years ago and has ended up today as an inselberg. Describe and explain the sequence of events and natural processes involved in the formation of Uluru. Include one diagram as part of your answer.

2   The surfaces of both Uluru and Kata Tjuta are smooth but also full of cracks, ridges, grooves and caves. Explain how these surface features have been produced.

3   How do the Anangu explain the shape and surface features of Uluru?

Uluru: between a rock and a sacred place — to climb or not to climb?

A proposed ban on climbing Uluru in Central Australia has sparked debate and controversy between tourists, traditional owners and political leaders.

What is Uluru?

Uluru is Australia's most recognisable land formation and a huge tourist attraction that is known for its distinctive shape and colour and for its many springs, waterholes, caves, ancient paintings and native flora and fauna.

Uluru and Kata Tjuta form the Uluru-Kata Tjuta National Park. In 1985, the Australian Government handed the park back to its traditional owners, the Anangu people. They, in turn, granted the Australian

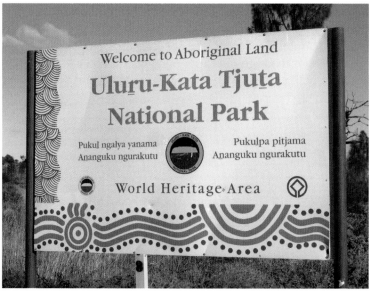

Figure 10

National Parks and Wildlife Service a 99-year lease, meaning the park is now jointly managed by the traditional owners and Parks Australia. For the Anangu, Uluru is a sacred place of great cultural significance.

Anangu are the traditional owners and guardians of Uluru. Their feelings are clearly and politely stated: *Wanyu Ulurunya tatintja wiyangku wantima* — please don't climb Uluru.

Are people allowed to climb Uluru?

Visitors have climbed Uluru for many years. However, since 1985, when the park was returned to the Anangu, awareness of Uluru's cultural significance has been recognised and has influenced the way the park is managed and used. The result is that the number of people choosing not to climb has increased: of the 350,000 annual visitors, only a third now make the climb. In 1990, the figure was 74 percent. This drop in the number climbing has not happened because of compulsion

Figure 11 Dangerous conditions can result in Uluru being closed to climbers.

but because of the decisions of people. More and more visitors recognise indigenous rights and beliefs and have responded in a positive way to the Anangu request that visitors do not climb on Uluru. Climbing the rock poses dangers from heat, steepness, strong winds and slippery surfaces when wet. The Anangu feel a responsibility for the safety of visitors and for this reason also they prefer that people do not climb the rock. However, Anangu do not prevent those who wish to climb from doing so.

Parks Australia recently released for debate a management plan for the Uluru-Kata Tjuta National Park

In reference to the Uluru climb, the plan stated: 'For visitor safety, cultural and environmental reasons, the director and the board will work towards closure of the climb.' Health and safety issues are listed as a constant challenge — 36 people are known to have died while attempting the climb, and park rangers often have to rescue those who have underestimated its steep and arduous nature.

Parks Australia research states that 98 percent of visitors would still come to the park if the climb were permanently closed.

What politicians say about prohibiting climbing
- Some local politicians have said that closing the climb cannot be allowed as it will halt 'one of the great tourism experiences in Australia ... Big Brother [the government] is coming to Uluru to slam the gate on an Australian tourism icon.'
- The Northern Territory Tourism Minister argued that individuals must be given the choice to climb or not to climb and added that 'that tourist numbers are dropping and closing the climb will not help matters'.
- Others support a permanent closure out of respect for indigenous law and culture. They say Uluru's beauty can be fully experienced without the need to climb.
- A Parks Australia director commented that the climb is not only dangerous but also out of step with the values the park seeks to uphold. 'It achieved world heritage listing for its outstanding natural features — and climbing is not that,' he says.

What other people and groups say
- 'The rock doesn't belong to anybody. People have a "right" to climb it. Look at the Grand Canyon in the USA, for example. It is more spectacular and more famous than Uluru, it is also culturally significant to local Indians, and it has no climbing restrictions.'

- 'If a cultural group wishes to attribute a creation story and sacred status to the rock, it is entitled to that viewpoint. If others, however, wish to climb the rock and enjoy what is truly a memorable experience, they should also be entitled to do so.'
- 'The proposed closure of the Uluru walking track shows a profound disrespect for the sacredness of world heritage sites for all humanity. Everyone should have the right to share in the wonder of climbing this breathtaking landform. Privileging the rights of traditional owners at the expense of what is arguably a human right to enjoy the world's natural resources, respectfully, is absurd. The rights of traditional owners need to be balanced reasonably against those of other stakeholders who also want to come close to its beauty.'

Figure 12 Tourists climbing Uluru.

Alternatives to climbing

Several tourism operators in the Uluru area offer Aboriginal guided tours without having to climb the rock. Tours include walking expeditions to explore rock formations and Aboriginal art sites around the base, escorted by local guides and an interpreter. Groups are also introduced to Dreamtime stories — the framework of Aboriginal mythology — as well as bush foods, traditional didgeridoo-playing, dot-painting and spear-throwing. A company tour manager said, "One hundred percent of our customers end up learning enough about Anangu culture during the tour that they simply choose not to climb out of their new-found understanding and respect for local culture.'

Tourism brings change

At Uluru, facilities had to be increased to keep up with the demands of tourists. Before 1983, tourists visiting Uluru were able to stay in a motel or camping park near the base of the rock. The airstrip was next to the motel and there was a dirt road from Alice Springs to Uluru.

By the early 1980s the number of tourists had become too large for the existing accommodation. Therefore a new complex of hotels and camping facilities was built. Called Yulara, the complex includes a luxury hotel, a more basic hotel and plenty of camping facilities.

The siting of Yulara was a problem. The complex was built some distance from the rock because of a combination of factors, such as the wishes of the local Aboriginal people, and the desire not to spoil the environment of the rock by building too close to it. This created a need for more efficient tourist transport to and from the rock. The larger numbers of tourists able to be accommodated created new demands for food supplies, sewerage engineering, water supplies and better roads. These demands have created new jobs in the area, but providing facilities was very expensive.

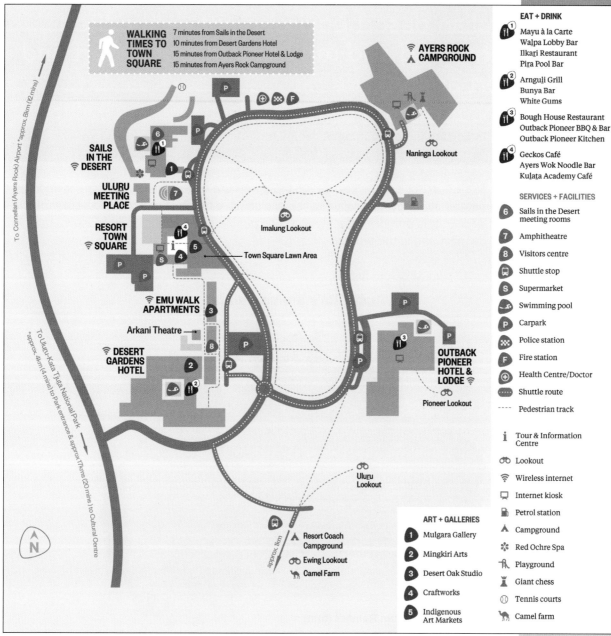

WALKING TIMES TO TOWN SQUARE
7 minutes from Sails in the Desert
10 minutes from Desert Gardens Hotel
15 minutes from Outback Pioneer Hotel & Lodge
15 minutes from Ayers Rock Campground

EAT + DRINK

1. Mayu à la Carte
   Walpa Lobby Bar
   Ilkari Restaurant
   Pira Pool Bar

2. Arnguli Grill
   Bunya Bar
   White Gums

3. Bough House Restaurant
   Outback Pioneer BBQ & Bar
   Outback Pioneer Kitchen

4. Geckos Café
   Ayers Wok Noodle Bar
   Kulata Academy Café

SERVICES + FACILITIES

6. Sails in the Desert meeting rooms
7. Amphitheatre
8. Visitors centre
S. Supermarket
P. Carpark
Police station
F. Fire station
Health Centre/Doctor
Shuttle route
Pedestrian track
i. Tour & Information Centre
Lookout
Wireless internet
Internet kiosk
Petrol station
Campground
Red Ochre Spa
Playground
Giant chess
Tennis courts
Camel farm

Shuttle stop
Swimming pool

ART + GALLERIES

1. Mulgara Gallery
2. Mingkiri Arts
3. Desert Oak Studio
4. Craftworks
5. Indigenous Art Markets

Figure 13 Ayers Rock-Yulara Tourist Resort.

Figure 14 Aerial view of the Ayers Rock-Yulara Tourist Resort.

Learning Activities

1 Make a list of points in favour of allowing the climbing of Uluru and a list of points against such climbs.

2 If you were on a visit to Uluru, would you make the climb? Give reasons for your answer.

3 Describe how the growth of tourism has brought change to the Uluru area.

The Australian desert environment — the 'Red Centre' a hot and arid region, a mix of desert and semi-desert areas

Important understanding: A natural environment is a distinctive part or area of the earth's surface with features that make it different from other areas.

Deserts cover a large part of Australia and the environment has many distinctive natural features (characteristics), including climate, landforms, vegetation and wildlife. It is an environment valued by people for many different reasons, both in the past and in the present.

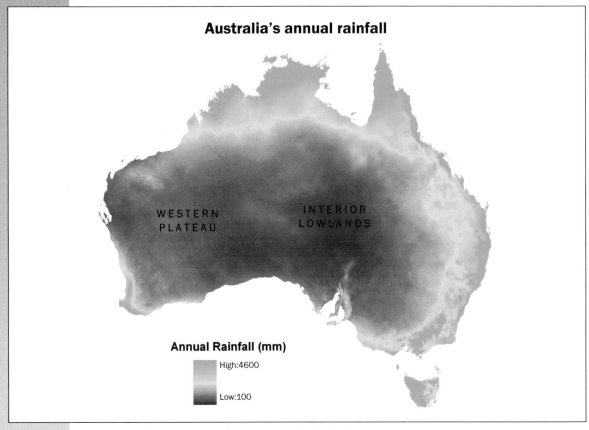

Australia's annual rainfall

WESTERN PLATEAU

INTERIOR LOWLANDS

Annual Rainfall (mm)

High:4600

Low:100

Figure 15 The Australian desert environment covers two thirds of the Australian land surface.

Desert and semi-desert areas

Places that get less than 250 mm of annual rainfall are described as 'desert' and areas with between 250 and 500 mm as 'semi-desert'. Australia is the driest inhabited continent in the world and has the largest southern hemisphere desert region. The desert environment covers more of Australia than any other environment type. Over one third of the continent is desert and another third is semi-desert. Desert and semi-desert areas combined cover 4–5 million sq km of the Australian land surface (Figure 15). Desert covers large parts of the Western Plateau and the Interior Lowlands. This area of desert is about 15 times larger than New Zealand and half the size of the USA.

A number of individual deserts are recognised by name within the desert zone (Figure 16). The largest desert, the Great Victoria Desert, is about one and a quarter times the size of New Zealand and the second largest, the Great Sandy Desert, is about the same size as New Zealand.

Aridity

Deserts are arid areas, which means they are places with a lack of moisture at ground level. The lack of moisture is caused by low amounts of rainfall but also by high rates of evaporation.

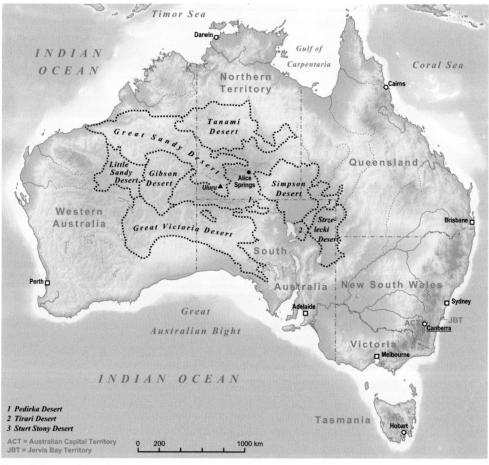

Figure 16 Deserts of Australia.

Characteristics of a desert environment

As well as being dry, the desert environment of Australia has other important and distinctive characteristics:

- Australia's desert environment makes up a large part of what is known as the 'Australian Outback'.
- The Australian desert is 'a warm-hot desert'. As well as little rain, the environment is one with high temperatures, low humidity and high evaporation rates. Daytime temperatures of over 40°C are common across the desert region.
- With little or no cloud cover, scorching days get replaced by very cold nights and temperatures falling below freezing occur each year.
- Nine hours of bright sunshine each day is typical of most of the desert region. Long periods of drought can be followed by short but heavy bursts of thundery rain, which can cause flash floods and lead to a sudden boost in the regeneration of plants and animals.
- Rainfall in the Australian deserts is erratic and unpredictable from year to year. Places can have less than half their long-term average for 10 years and then have all their annual average rainfall come in a single month or, on rare occasions, all fall in a single day. Periods of heavier rain often come when moist tropical air that normally sits over northern Australia pushes further south and brings wetter conditions into the desert area.
- Thunderstorms and dust storms are common. The thunderstorms are often 'dry' storms with most or all rain evaporating before reaching the ground. Some places have more days of thunder each year than they have days of rain.

- Due to the low level of rainfall in the arid zone, the rivers and drainage channels are usually dry and flow only occasionally after rain storms. When they do flow, the water either floods out onto the open flat country and ends up in salt pans, or soaks into the sand. Water becomes trapped beneath the sand but still provides moisture to plants and animals during dry times.
- Desert landforms have variety: large sand-covered areas forming plains or dunes, areas of stone and bare rock and many salt lakes. The land surface is mostly of low elevation but there are several upland areas in the central and western desert areas.
- The environment has been described as 'moderately desert' because this 'desert area' of Australia receives more rainfall and contains more moisture than desert areas like the Sahara (Africa) and Atacama (South America). It has also been described as having 'boom-bust' cycles: in wetter years desert grasses can be waist high; during dry times the plants shrivel and the desert land surface becomes bare sand.
- Plants, shrubs and trees adapted to the dry, hot conditions and able to survive in soils that are mostly infertile provide a thin vegetation cover across much of the desert land surface. The large amount of vegetation present is a feature not found in desert areas in other continents.
- The desert has been the traditional Aboriginal homeland for more than 50,000 years. Today, there are about 1300 Aboriginal communities with their homes in desert lands of Australia. These communities are small, with many having fewer than 50 people. They exist in some of the most remote and least disturbed country in Australia.
- Only 3 percent (about 600,000 people) of the Australian population live in the desert areas, which cover 70 percent of the continent. Some of the desert is totally uninhabited and all of it has a low density of population.
- The pastoral farming (sheep and cattle) industry has traditionally been the main economic activity found in the desert environment. Huge 'farms' (stations) are located within the semi-desert area. In some places today, tourism and mining are economically more important than farming.
- Increasing numbers of tourists are visiting the desert area looking for something different — the stunning landscape, the unique plants and animals — and to experience the open spaces, remoteness and quietness of the environment.

Climate

Climate is the most obvious and distinctive characteristic of desert environments. Deserts are arid places with low rainfall and high evaporation. The desert environment area of Australia is made up of both arid and semi-arid areas (see Figure 17).

Putting a boundary on the Australian desert environment

Australia's desert environment does not have an abrupt boundary where it suddenly stops. There is a transition into semi-desert and then pure desert conditions from wetter surrounding areas. Use of the 500 mm annual isohyet (an isohyet is a line marking places with the same rainfall total) is one way of putting a specific boundary on the area. Inside this line, places all have annual rainfall totals of less than 500 mm. These are the desert areas of Australia (see Figures 17 and 18).

The area is huge — 3000 km from west to east and 1500 km from south to north. By road and driving without any stops, it would take 19 hours to travel from Port Augusta in the south to Tennant Creek in the north (Figure 18). Travelling west to east from Geraldton to Bourke would take 40 hours. People making such trips would allow several days to a week to complete the journey. Flying time gives another indication of the huge size of the desert environment. From Perth to Alice Springs in the centre of the desert area, the flying time is three hours; from Adelaide to Alice Springs, the flight is two hours.

 ISBN: 9780170233316

The Australian desert environment area

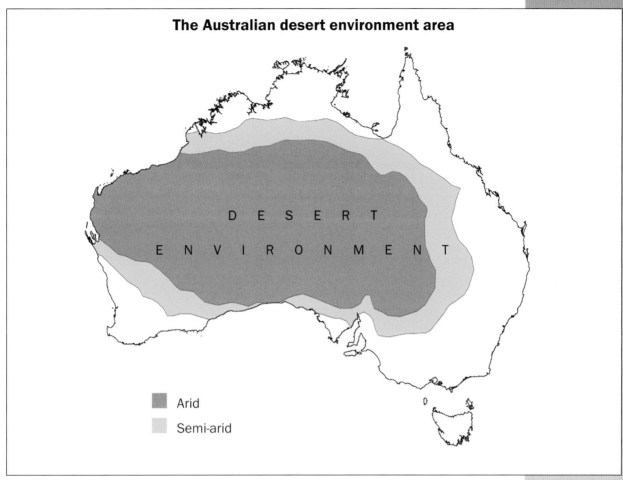

Figure 17

Australia's annual rainfall

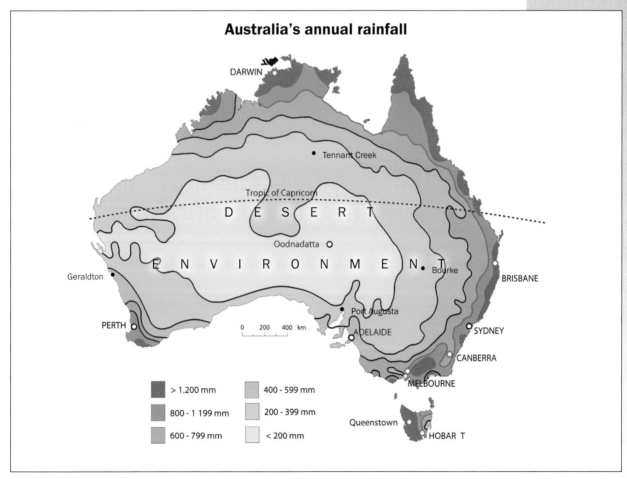

Figure 18

1 a Using Figure 16 as a guide, either name on the map or number on the map and key the names of the four largest deserts: Great Victoria, Great Sandy, Tanami, Simpson.

b Refer to Figure 17. On a large outline map of Australia, show the boundary of the environment area. Inside this boundary, shade to show the arid and semi-arid parts that make up the environment.

2 'The desert environment is huge in size.' List five facts to support this statement. For example:

i Desert covers two thirds of the land area of Australia.

3 The Australian desert environment is an arid environment.

a What does the term 'arid' mean?

b What two things in combination make a place arid?

4 Figure 15 highlights that there is a transition from desert to non-desert areas rather than there being an abrupt and precise boundary line.

a How does the map in Figure 15 show this transition idea (what mapping/shading method has been used on the map)?

b Why in the real world would there be a transition from desert to non-desert rather than there being a precise boundary?

5 Describe the rainfall pattern shown in Figure 18.

6 Draw a star diagram to show eight important features of Australia's desert environment. Two of these should be cultural features. Use different colours for the natural and cultural features. One example of a natural feature has been included as an example.

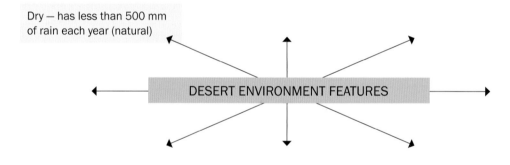

Dry — has less than 500 mm of rain each year (natural)

DESERT ENVIRONMENT FEATURES

Characteristics of the Australian desert environment — natural elements with a human contribution

The elements or characteristics of the Australian desert environment are shown in Figure 19.

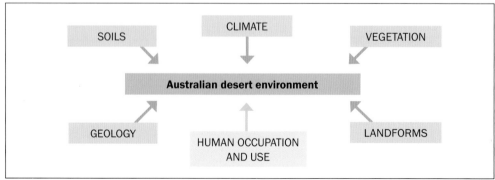

Figure 19 The elements that make up the environment.

 ISBN: 9780170233316

Elements, processes and interactions within the Australian desert environment

Important understanding: Elements and natural processes interact within the Australian desert environment. These interactions have helped to shape and create a distinctive environment. Movement and change happens all the time within the desert environment — the environment is 'dynamic'.

Elements and characteristics

Elements are the parts of something. Elements of a natural environment are the landforms, geology, climate, soils and vegetation — these are the parts that together make up the environment. Elements can also be referred to as the characteristics or features of the environment. The elements themselves are made up of smaller elements, called sub-elements. Climate, for example, is an element and is made up of sub-elements such as temperature, precipitation, sunshine, winds and humidity.

Processes

Processes involve a sequence of actions that shape and change the environment. In the Australian desert environment, wind (aeolian) processes, river (fluvial) processes, weathering and climate processes are examples of important natural processes that operate. Processes involve a number of elements. For example, the nature and speed with which wind erosion takes place is determined by wind strength and direction and the nature of the rocks and the ground surface cover across which the wind blows. Processes of the past have had a big influence of the landforms of the present day.

Interaction

Interaction means things being linked together and affecting each other. The landscape of the Australian desert is the result of interaction between natural elements and natural processes. Interaction also causes the landscape to be continually changing. Processes operate all the time and as a result the landscape is always changing, both in the short term and long term. Process operation and landscape change are the reasons the environment can be called 'dynamic' — it is full of movement and change.

Learning Activity

1. a   Copy Figure 19 and make a key for the two different-coloured arrows (blue and green) used in the diagram.

   b   How many *natural* elements make up Australian desert environment?

   c   What does the term 'sub-element' mean? Give an example as part of your explanation.

   d   Which natural process is controlled by wind speed and type of land surface cover?

   e   The desert environment (like other environments) can be described as 'dynamic'. What does this mean and what causes it?

Climate of the Australian desert environment

Important understanding: Climate is the most important feature of the Australian desert environment. 'Dryness/Aridity' defines the environment. Climatic processes that lead to low rainfall and high rates of evaporation are the cause of the aridity. Desert weather (day by day and week by week conditions) is unpredictable and can change rapidly.

The desert environment covers most of central and western Australia. This is a dry area, with fewer than 40 days when rain falls each year (Figure 20) and annual rainfall totals of less than 500 mm (Figures 17 and 18). Southeast of Alice Springs around Oodnadatta and Lake Eyre there are fewer than 10 rain days each year and annual totals are between 100 and 140 mm each year. This is the driest part of Australia.

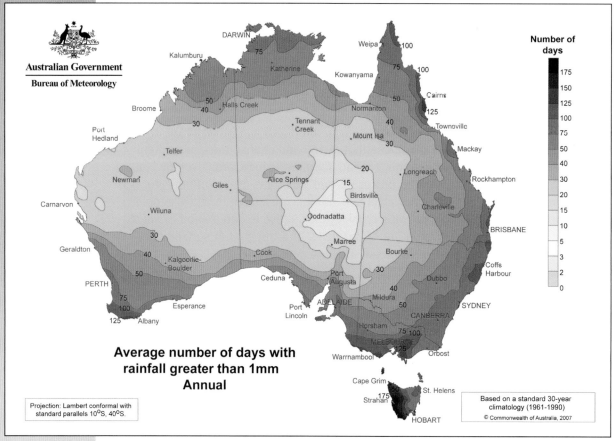

Figure 20 Number of days with rain.

High temperatures and dry air mean that any rain that falls in the desert is quickly evaporated. Little water is left to sink into the ground or flow across the ground surface. There is the potential for 5–10 times more evaporation to take place than there is water available.

Rainfall amounts and the number of days with rain decrease towards the centre of the desert environment area. Temperatures are 'hot', well above 35 °C during the day across large parts of the central and northern desert areas, but these decrease southwards as distance from the tropics increases. The northern desert areas are 10 °C hotter than areas on the southern margins of the desert. Even though the days are hot in the desert, the clear skies and lack of cloud cover mean the heat escapes quickly when the sun goes down and temperatures can drop rapidly to be 15–20 °C cooler than the daytime temperature. Frosts are common in the desert at night. The difference between daytime and night-time temperatures is called the 'diurnal temperature range'. Australia's desert environment has a large diurnal temperature range.

Local variations in climate also occur. Alice Springs (Figures 21 and 22) is 'less hot and less dry' than the surrounding plains because of its higher elevation (600 metres) reducing temperatures and the orographic influence (rising and cooling air leading to cloud formation and some rainfall) of the nearby MacDonnell mountain range.

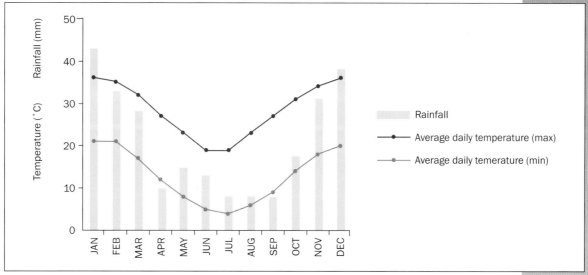

Figure 21 Climograph for Alice Springs.

	Annual rainfall total (mm)	Number of days with rain each year (< 1 mm)	Max daily temp summer Jan (°C)	Min daily temp summer Jan (°C)	Max daily temp winter July (°C)	Min daily temp winter July (°C)	Number of days with temp above 35°C	Annual hours of sunshine
Tennant Creek: northern semi-desert zone	471	49	37	25	24	12	124	3577
Alice Springs: central desert zone	?	30	?	?	?	?	88	3504
Oodnadatta – Lake Eyre: the driest desert area	176	22	38	23	20	6	91	3650
Auckland, NZ – for comparison	1240	136	23	16	14	8	0	2060

Figure 22 Australia desert locations climate features.

Why is the climate dry?

The brief reason is that the part of Australia that is desert is located in an area of subtropical high atmospheric pressure, which results in lots of dry air and cloudless days.

Australia is the world's driest inhabited continent (Antarctica is drier) and contains the largest desert region in the southern hemisphere. The main reason so much of Australia is desert is because of the location of the continent. Many of the world's greatest deserts are located around latitudes 20–30° north or south of the equator. In the northern hemisphere, the Sahara Desert and the deserts of the Middle East and the southern USA are in this latitude band; in the southern hemisphere the Australian desert is within a similar band (Figure 24). In these latitude bands, climate is controlled by global air circulation patterns. A sequence of events takes place (Figure 23):
- Hot moist air rises at the equator and begins a movement northwards and southwards towards the poles.
- The air cools as it rises. This causes moisture in the air to condense and form rain clouds above the equatorial areas. Heavy rain falls in these places.

- The now dry air continues to travel high in the atmosphere away from the equator. This air begins to sink back towards the earth's surface in the subtropical areas around 20–30° north and south of the equator.
- These 20–30° latitude areas get a double inflow of dry descending air because they are also affected by descending air that has been flowing towards the equator from temperate (cooler) areas nearer to the poles. The two flows of descending air pushing down onto the earth's surface lead to the formation of areas of strong high pressure in the subtropical latitudes (the 20–30° latitude bands).
- The descending air masses are dry and warm up as they descend. They contain little moisture to form clouds but instead bring clear skies, lots of daytime sunshine and low rainfall. The clear sunny skies result in strong daytime heating, low night temperatures and high rates of evaporation.
- Dry arid areas are the result — one of these areas is the Australian desert area.

In addition to being located in the subtropical high-pressure zone, the interior of Australia is kept dry by three other factors (Figure 23):

1 Moist air travelling from the Pacific towards Australia brings regular rain to the east coasts of Queensland and New South Wales. The Great Dividing Range blocks the inland movement of this moist air. A rain-shadow effect occurs inland of the mountains and this adds to the dryness of the eastern part of the desert.

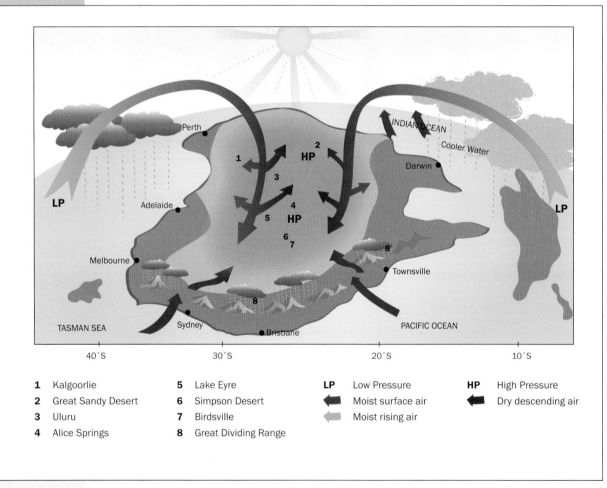

1	Kalgoorlie	**5**	Lake Eyre	**LP**	Low Pressure	**HP**	High Pressure
2	Great Sandy Desert	**6**	Simpson Desert		Moist surface air		Dry descending air
3	Uluru	**7**	Birdsville		Moist rising air		
4	Alice Springs	**8**	Great Dividing Range				

Figure 23 Why Australia is a dry continent.

2 There are no large high mountainous areas or large water sources in the interior of Australia. This means there are no mountains to force air upwards and create orographic (relief) rainfall and no sources of water to replenish moisture in the dry air as it approaches the centre of Australia. Dry air across the interior is the result.

3 Sea surface temperatures also impact on Australian rainfall. Cool ocean water offshore of northwest Australia, and warmer waters further west in the Indian Ocean, causes a frequent flow of air westwards and away from Australia. This ocean temperature condition restricts the development of onshore airflows from the Indian Ocean that would bring rain to the western side of the continent.

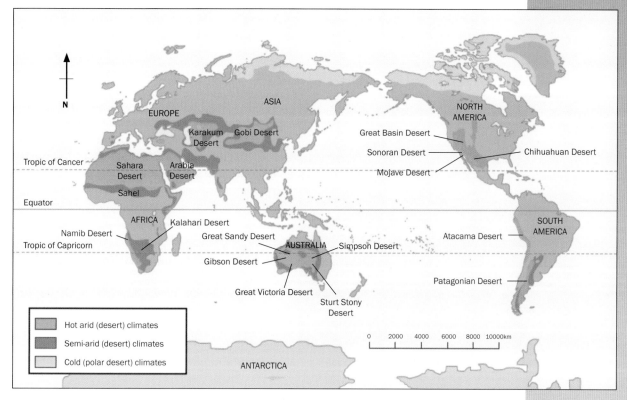

Figure 24 Australian deserts are part of the global pattern of deserts located around 20–30 degrees north and south of the equator.

Learning Activities

1 Refer to Figures 18, 20 and 23.

   a   **i**   Describe the location of Oodnadatta.

      ii   What is distinctive about the climate of this Oodnadatta area?

   b   Give examples of how the desert climate varies from place to place (shows spatial variation).

   c   Explain why these variations occur.

   d   Make a bullet point list of four distinctive features of Australia's desert area climate.

2 **a**   Refer to Figures 21 and 22. Following the format of Figure 22, draw a climate summary table but show information for just Alice Springs and Auckland. You need to calculate the missing data for Alice Springs (annual rainfall and daily temperatures) from the information given in the climate graph in Figure 21.

   b   In a short paragraph, compare and contrast the climate of Alice Springs with that of Auckland.

3 Refer to pages 82–87. Write a long paragraph answer explaining why so much of Australia is dry. Include each of these five factors in your answer:
- global (latitudinal location)
- high pressure
- rain shadow
- interior landform features
- sea surface temperatures.

88

Vegetation of the desert environment

Important understanding: About half the area of the Australian desert environment has a vegetation cover. This vegetation is found in places where moisture is available to support plant growth. Vegetation is adapted to survive both arid conditions and poor soils.

A tourist guide about the Australian desert environment includes this description:

'Many people hear the word "desert" and expect endless sand dunes, or barren stony plains without vegetation. The Great Victoria Desert [the largest individual Australian desert — see Figure 16] looks nothing like this. It's called a desert because there is little rain, not because it is dead or boring.

Figure 25 Eucalyptus (gum) trees with deep roots surviving on a rocky desert outcrop.

The amount of vegetation may surprise you. Australia has always been a dry continent, and the plants are well adapted to living with very little water. Not only marble gums, mulga and spinifex grass thrive here. You will also find a huge variety of shrubs and smaller plants.

When it does rain the transformation is total. The desert bursts into bloom seemingly overnight. Fields of wildflowers, accentuated with flowering grevilleas and acacias, yellows, whites and mauves against the red sands. The sight of the blooming desert is something you will never forget ...'

The amount of desert with permanent vegetation cover is higher in the Australian desert environment than in any other major global desert. This is because the Australian desert has slightly more rainfall and therefore more moisture available to support plant growth than in other deserts.

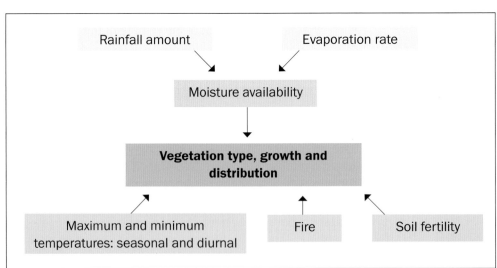

Figure 26 Factors that influence desert vegetation.

The plants have evolved to conserve and efficiently use any water that becomes available to them. The most common plant communities in the desert are triodia (spinifex) grasslands, saltbush, bluebush, mulga shrublands and mallee woodlands.

Desert plants are clever survivors in difficult conditions. They survive in two ways: through drought tolerance and drought avoidance.

 ISBN: 9780170233316

i **Drought tolerance** Plants known as xerophytes have a structure that allows them to find water, store water and limit their water use. They have tough leaves with a waxy or hairy surface that minimises surface water loss through their leaves (transpiration losses are low). They store water in succulent leaves and stems or in fleshy underground tubers. They also have deep root systems to access deep underground water supplies or a shallow and spreading root system that lets them take up moisture quickly after rainfall. Triodia (spinifex) species are xerophytic.

ii **Drought avoidance** Desert plants can avoid death through moisture stress by remaining dormant during dry periods and then 'springing to life' when water is suddenly available. The name for plants of this type is 'ephemeral'.

Ephemerals live for only a short time and appear only when there is a lot of moisture after heavy rains. The plants respond rapidly to rain — germinating, growing, flowering and setting seed all within a very short period of time while the desert soil remains moist. For a short time, ephemeral plants like paper daisies and Sturt's desert pea can produce a carpet of colour across the desert. As drought returns, these plants wilt and disintegrate, scattering their seeds in huge numbers across the desert waiting for the next heavy rain to set their germination in motion.

Figure 27 Sturt's desert pea plant.

The small seeds of Sturt's desert pea can germinate after many years of lying dormant. The seeds have a hard coating, which protects them from harsh arid environments until the next rainfall. Once they germinate after heavy rain, the seedlings quickly establish a deep taproot to seek out below-ground moisture. Sturt's desert pea produces distinctive red flowers that set seed in a very short time.

Mulga (low woodland and shrublands) areas are located where rainfall is slightly higher than in the spinifex areas and also where fire has been less frequent.

Acacia shrubs are common Australian desert environment plants. Acacia shrubs, like many other desert plants, have more of their growth below ground than above ground. Their deep root systems search out and tap into underground water supplies.

Figure 28 Mulga (acacia) trees with triodia surface grasses.

Different local environment conditions result in other vegetation types being found in specific desert locations where adaptations to the local environment conditions take place:

a In 'usually dry' river channels which flow with water only after heavy rain, there are below-surface moisture supplies. Tree species like river red gums (eucalyptus) tap into this supply with deep root systems. These trees often grow in a linear pattern along the channels.

b Wetland vegetation of ferns and sedges are found in the few areas with permanent freshwater supplies. These water supplies are found in spring-fed rock pools and waterholes.

c In salt lake areas like Lake Eyre, there are plants that can survive both high salt concentrations in the soil as well as the very hot and dry conditions that exist for most of the time.

Triodia desert grasslands

(Note that these are usually called spinifex grasslands in Australia. Triodia-spinifex is not the same as coastal area spinifex.)

These plants thrive on the poorest, most arid soils in Australia. The plant has helped bind the desert sands and prevented the Australian desert environment from becoming a Sahara-like world of bare, shifting sand.

Triodia-spinifex are hard-leaved grasses that form hummocks of vegetation. The hard, waxy leaves roll tightly into vertical spikes, minimising exposure to the midday sun to prevent water loss. Spinifex roots go down a long way — as much as 10 metres — and provide each clump with its own water supply from deep underground.

These grasses grow across much of the desert area where ground conditions are arid and soil fertility low.

Figure 29 Spinifex (triodia) hummocks in Great Victoria Desert with acacia bushes in the background.

Spinifex grasslands are the most widespread vegetation type in Australia, covering 22 percent of the continent. They are the main plant type found across all areas of the Great Sandy Desert and are common in moister low areas between the sand dune ridges of the Simpson and Great Victoria Deserts.

The plant grows in many habitats — sand plains and dunes, and on rocky hills and mountain range slopes.

Fires over a long time period have resulted in the presence of various fire-tolerant shrubs like mallee eucalypts in among the spinifex grasslands.

Spinifex tussocks are very flammable and burn easily because they are formed of clumps of matchstick-sized blades of dry grass filled with flammable resins with lots of air space between the blades. On a scale of 1 (lowest) to 10 (highest) for flammability, spinifex scores a 9. A small lightning strike or one match can start a raging fire. Spinifex grasses give off dark smoke, which can be seen from far away. Aborigines used fire to send long-distance smoke

Figure 30 Spinifex (triodia) burning.

signals, to manage habitats and keep terrain open, as well as to help capture animals for food.

 ISBN: 9780170233316

Learning Activities

1   Copy and match up the information in column A with that in column B.

A Terms	B Definitions
Triodia desert grass	Bright-coloured and ephemeral plant that includes the name of a desert
Eucalyptus gums and mulga bushes	Spiky plant with deep roots with a common name of spinifex
Sturt's desert pea	Place where vegetation grows in a linear pattern
Lake Eyre	A natural cause of desert fires
Dried-up riverbeds	A place where plants need to be both drought and salt resistant
Lightning	Plants adapted to survive in an arid environment
Xerophytic	Two larger types of desert vegetation

2   **a**   What makes it difficult for plants to survive in the desert environment? Show your answer as a bullet point list or in a star diagram.

   b   Explain how plants like those shown in Figures 27–30 are able to survive and live in the desert environment. Write your answer as a long paragraph. Include specific plant names within your answer.

Soils and geology of the desert environment

Important understanding: The soils of the Australian desert environment are very old, shallow and low in nutrients. They are low-quality soils and also very fragile and easily eroded.

Soils form from the weathering of rock (break-up of rock into small particles) which becomes mixed with decaying plant material (humus) and animal excrement. Soils in the Australian desert environment are infertile with low levels of nitrogen and phosphorus — less than half that found in other arid regions of the world. They also have little humus content.

The rocks beneath the surface and on the surface are ancient — millions of years old. Soils have been developing from these rocks for a long time period. In many parts of the world, present-day soils are young because glacier movement across them during ice ages of the last 50,000 years stripped away all their soil. New young soils have since developed from fresh rock surfaces and from material deposited when the glaciers and ice retreated. Australia was not covered by these ice sheets and so the modern-day soils are ones that have formed on land surfaces where soil development has gone on continuously for many millions of years.

For most of the time of the soil formation, the climate of Australia has been much wetter than it is today. The soils have had their nutrients washed away (leached) during the long time period of their formation in these wetter conditions as well as being weathered (broken up) to great depths.

The soils of the desert environment area arid zones fall into three categories:
a   areas of bare rock and stones where smaller particles have been removed by wind and water

b deep red-brown coloured sand which covers dunes and plains to a depth of several metres

c red earths which have formed from rock weathering that has left iron oxides in the soil giving them the red colouring. These soils are low in humus and often have a high salt content.

Case Study 2

The Great Artesian Basin (GAB) – geology and climate combine to produce a world-famous feature

Underneath a large part of the interior lowlands and eastern part of the Australian desert environment is the Great Artesian Basin (Figure 31). This is one of the largest underground water reservoirs in the world.

Water from underground rises to the desert surface at natural springs where faults and fractures in the overlying rocks allow the pressured underground water to force its way to the surface. Springs also occur where the water-carrying rocks reach the ground surface at the edges of the basin (Figure 32). There are more than 600 natural surface springs spread across the basin. The water reaching the desert surface is described as 'drinkable fresh water' but frequently has a high mineral and salt content. In the places where the water comes from deep underground, it reaches the surface as hot water with temperatures of between 30 and 50°C and can be as high as 100°C.

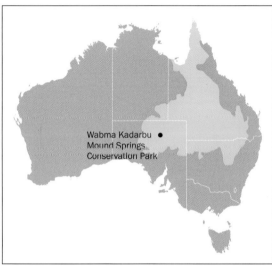

Figure 31 Location and size of the GAB (blue area).

A lot of the water flowing to the surface today fed into the basin millions of years ago when the climate was wetter, but water continues to feed into and resupply the basin today from the Great Dividing Range (Figure 32).

The term 'artesian water' is used to describe water trapped underground between rock layers, which rises upwards towards the ground surface because of natural pressure.

Figure 32 A natural artesian spring in the Wabma Kadarbu Mound Springs Conservation Park on the southern edge of the Great Artesian Basin in South Australia.

 ISBN: 9780170233316

Case Study 2

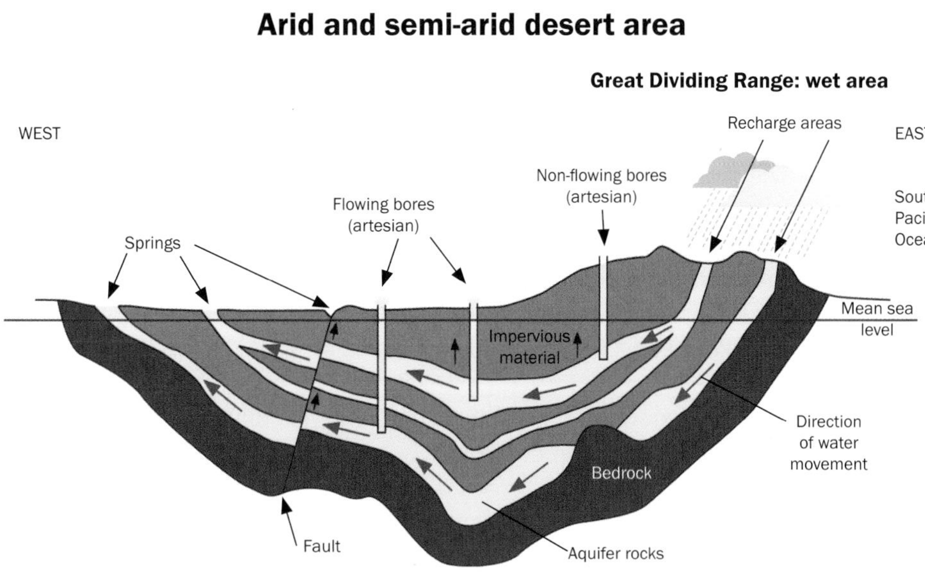

Arid and semi-arid desert area

Great Dividing Range: wet area

Figure 33 How the GAB works — a combination of geological and climate processes.

Bedrock	Oldest rock — impermeable and will not let water pass through.
Aquifer	Permeable sandstone rocks — allow water to pass through them and water becomes stored in pore spaces between the sand grains within the rock. These are the rocks that hold the underground water supplies.
Direction of water movement	The water travels very slow speeds of 1–5 metres per year through the rocks from the Great Dividing Range recharge area.
Impervious material	Younger surface rocks made of siltstones and mudstones which cap and trap the water underground.

Learning Activities

1. Make a list of reasons why the soils of the desert environment area are soils of low quality.

2. a Describe the location of the Wabma Kadarbu Mound Springs Conservation Park.

   b Draw an annotated sketch of the desert oasis in Figure 32. Attach these labels to the sketch:
   • Artesian spring
   • Wetland ferns and sedges
   • Xerophytic vegetation
   • Desert plain.

3. a Explain how the Great Artesian Basin works — describe and explain how water moves into and through the basin. Mention 'geology' and 'climate' in your answer.

   b Why is the GAB important to the eastern part of the Australian desert environment?

Landforms of the desert environment

Important understanding: The Australian desert environment is made up of a variety of landforms. These landforms are the result of weathering, wind and water processes operating over a long period of time.

The rocks that make up the desert environment are hundreds of millions of years old. During this time, mountains have been worn down. Because Australia is far away from tectonic plate boundaries, there has been little recent uplift of land or volcanic activity to reform mountains and re-create higher land. Most of the desert lands of Australia are low lying with a flat or undulating surface. Rocky outcrops like Uluru and Kata Tjuta stand out above the plateau and plains areas.

Surface desert landforms show the influence of weathering, wind (aeolian) processes and water (fluvial) processes:

a Weathering

i Mechanical weathering is a process that causes rock to gradually break up into smaller pieces. The word 'weathering' is used because it is the action of the weather that causes the rock to break up. In deserts, the sun beating down onto bare rocks can lead to rock surface temperatures reaching 75°C. At night, air temperatures cool rapidly. This results in day-night temperature differences of 25–30°C. (The day-night temperature difference is called the diurnal temperature range — in desert areas, this range is large.) The rock surfaces cool down as well. This great change in rock temperature causes daytime expansion and night-time contraction of the rock surface layers. This stresses the rocks and leads to their cracking, shattering and gradual disintegration. This process is called mechanical weathering. Mechanical weathering is sped up if just small amounts of water are present from things like fog, or from underground moisture moving upwards to the surface. Big changes in temperature combined with the presence of moisture cause the maximum amount of mechanical weathering.

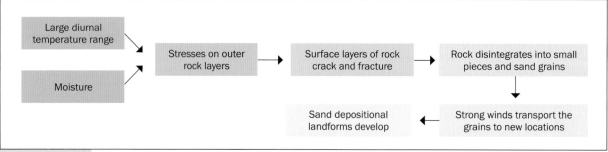

Figure 34 Mechanical weathering, wind transportation and wind deposition.

ii Chemical weathering results from crystal growth on rock surfaces or in cracks in the rocks. The growing crystals force the rocks apart. The crystals grow from salt and other minerals left after evaporation when rivers or lakes dry up following heavy rain storms. Break-up of the rock into smaller and smaller pieces through weathering reduces the rock to grains of sand that are small enough to be carried and deposited by the wind.

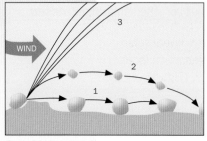

Figure 35 Moving sand.

b Wind (aeolian) processes. Fine, dry and loose desert sand can be easily transported and then deposited by the wind. This wind action creates many distinctive desert landform features. Depending on the sand grain size and weight, and on the strength of the wind, the sand can be moved in one of three ways: surface creep, saltation, suspension.

 ISBN: 9780170233316

Moving sand

LEAST MOVEMENT ·· ▶ MOST MOVEMENT

1 **Surface creep** Larger and heavier grains of sand are rolled across the ground surface, usually in a stop-start way as wind strength rises and falls.	**2** **Saltation** Sand moves in a hopping motion. It gets picked up by a wind gust, transported in the air and then deposited when the wind strength dies away. The next strong gust picks up the sand again, moves it and then drops it in a repeat of the first cycle.	**3** **Suspension** The smallest and lightest sand grains are picked up by the wind and can be transported hundreds of kilometres before being deposited back on the desert floor.

Figure 36 Three types of sand transportation.

In parts of the Australian desert environment, loose sand has been carried away by the wind leaving behind a desert surface covered by a layer of stones called gibbers. (The name 'gibber' comes from the Aboriginal name for stone.) The area covered in these stones is called a gibber plain (Figures 37 and 38). In other desert areas, like the Sahara, stony deserts are called 'regs'.

Figure 37 Gibber desert surface (rock covered).

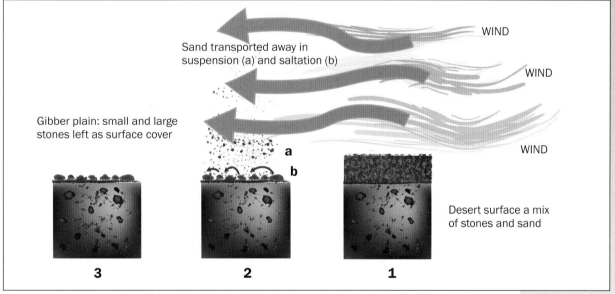

Figure 38 Gibber plain formation.

Sand desert areas (called erg deserts) These are areas covered by a layer of sand that has come either directly from weathering or from wind deposition. Across the Australian desert environment there are large generally flat or gently undulating sand areas which are often vegetation covered, especially in the semi-arid areas. In many of these sand desert areas, there are sand dunes. Crescent-shaped dunes called barchans are not common in the Australian desert environment. Instead, linear (also called longitudinal or seif) dunes, clustered together in long parallel ridges, form large sand field areas and cover hundreds of kilometres of the desert surface (Figures 41, 42 and 43). The Australian desert area has some of the largest areas of linear dunes in the world. These dunes are another feature produced by wind transport followed by wind deposition.

c **Water (fluvial) processes** Although the desert environment is dry and arid, water still has an influence on desert landforms and had a much greater influence in the past when the modern desert area had a wetter climate. Some of the most prominent landforms like Uluru were formed long ago when conditions were wetter, and modern desert conditions have only modified them rather than created them. Today, natural springs are found in some places where water from the Great Artesian Basin reaches the surface (pages 92–93). Although under the present climate conditions rain is infrequent and unpredictable, the desert does still get periods of heavy rain and storms. When these occur, dry river beds (wadis) and river valleys fill with water and rivers flow across the desert surface. This flow maintains the river channels and ends with the water accumulating and forming shallow lakes in low-lying basin areas. The rivers and lakes are ephemeral (short lived) because water is either evaporated or quickly sinks into the ground once the rain stops. The evaporation leaves behind salt deposits on the river and lake beds. The dry salt lake beds and the salt lakes are called playas. Lake Eyre is a world-famous example of such a feature (pages 104–106).

Great Victoria Desert

Has a mix of small sandhills and desert pavements where wind and water have eroded away the fine sands and left flat areas of bare rock, or rock covered by small and closely packed pebbles called gibbers (Figures 37 and 38). Many small playa (salt) lakes (Figures 50 and 51).

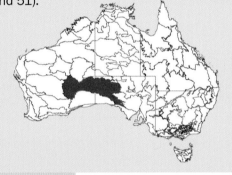

Great Sandy Desert and Simpson Desert

These are ergs (sand deserts) with red sandy plains and large areas of linear sand dunes (Figures 41 and 42). The deserts contain many playa lakes including Lake Eyre (Simpson Desert) and inselbergs like Uluru (Great Sandy Desert).

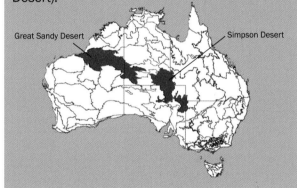

Gibson Desert

Has many gravel-covered gibber areas with an uneven covering of desert grasses. Other parts of the Gibson are formed of bare red sand plains and dunefields interrupted by low rocky/gravelly ridges and small playa lakes.

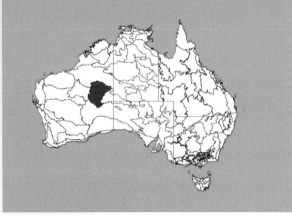

Sturt Stony Desert

A reg desert (i.e. a desert with a surface cover of bare rock and stones) with a covering of small to large-size stones (Figures 37 and 38). The stony surface is the result of weathering breaking down the ancient sandstone rocks into hard rock fragments which end up covering the surface of the desert.

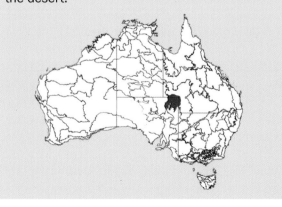

Figure 39 Australian desert environment landform features.

 ISBN: 9780170233316

Natural elements, processes and interactions within the Australian desert environment — a summary

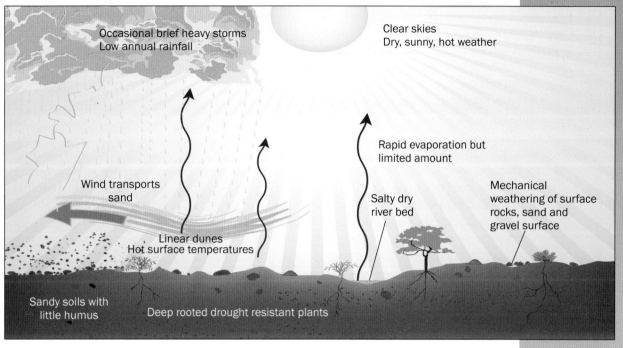

Figure 40 Elements, processes and interactions within the Australian desert environment — climate, vegetation and soil linkages.

Learning Activities

1 Refer to Figure 39.

a Make a list of the five deserts named in Figure 39 in alphabetical order.

b What is a desert surface covered in small rocks and stones called?

c Which landform feature could be sometimes wet and salty and at other times dry and crusty?

d Name the two erg deserts with linear sand dunes.

e Name an example of a rocky reg desert.

2 **a** Make a copy of either Figure 35 (wind transport) or Figure 38 (gibber plain formation). In no more than 30 words, explain how the processes operate (works) in the diagram you have copied.

b Write a short two-part essay (include illustrations in your essay):

i Describe and explain how the processes of weathering, wind transportation and wind deposition operate and are connected.

ii Explain how two named desert landforms result from the operation of these three processes.

3 **a** Copy Figure 40. Label or number on your diagram:

- two examples of elements (e.g. linear dunes)
- two examples of processes (e.g. evaporation)
- two examples of interaction (e.g. vegetation and soil linkages).

b Use your knowledge and understanding of the Australian desert environment to complete this task.

i Choose one of the elements you have identified in your diagram (1a). Write a paragraph *giving detailed information* about this element.

ii Choose one of the processes you have identified and write a detailed paragraph *explaining how* this process operates.

iii Choose one of the interactions you have identified and write a detailed paragraph that *describes and explains* the interaction that takes place.

The Simpson Desert

The Simpson Desert has three very distinctive features:
- Beneath the desert is the Great Artesian Basin. Natural springs on the desert surface result from the presence of the artesian water (pages 92–93).
- The Simpson Desert contains a huge system of parallel longitudinal (seif) sand dunes — sometimes said to be the longest in the world.
- Lake Eyre, the largest salt lake (playa) in Australia, is located within the Simpson desert (page 104).

The Simpson Desert is a sand desert that contains large areas of longitudinal sand dunes formed of quartz sand grains. There are over 1000 of these dunes and they run in a SSE-NNW direction. The dunes have been formed by winds blowing generally from south to north but from two different angles. These winds have reshaped and deposited sand into a series of long parallel ridges (Figure 41).

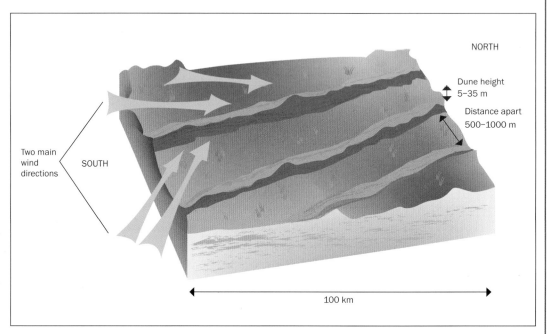

Figure 41 Longitudinal dunes.

Figure 42 Big Red (Nappanerica) longitudinal sand dune on the eastern edge of the Simpson Desert near Birdsville in southwest Queensland.

The sand dune ridges are long with some over 200 km in length. Most of these dunes today are fixed, held in position by vegetation (sandhill canegrass on the dune tops and sides and spinifex in the low areas between the dune ridges); a few remain as bare sand and travel. The dunes have an average height of 20 metres and are about 500 metres apart. Weathering of the sand grains releases iron oxide and this gives the dunes their brown-red colouring. The largest and most famous dune, Nappanerica (commonly called Big Red) is 40 metres in height (Figure 42).

Case Study 3

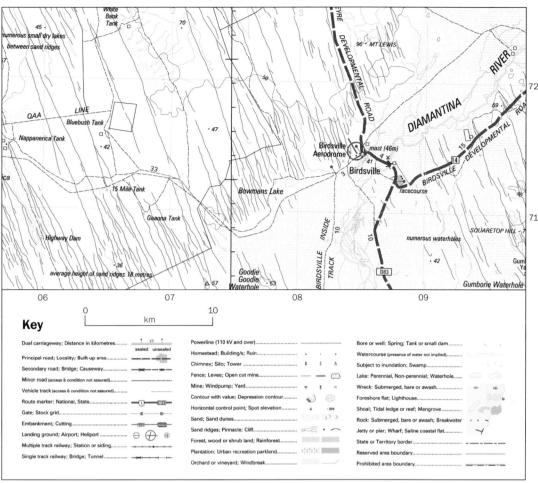

Figure 43 Topographic map of the Birdsville area of the Simpson Desert.

How the Simpson Desert has formed and changed over time

270 MILLION YEARS IN THE MAKING:

CHANGING SEA LEVELS + LAND MOVEMENTS + SEDIMENTARY BASIN FORMATION + DEPOSITION + CLIMATE CHANGES = MODERN-DAY SIMPSON DESERT

- Around 270 million years ago the area that is now the Simpson Desert was glaciated, and after this covered by a rising sea. When the sea level fell again, the area was covered by many shallow freshwater lakes. Organic material was deposited on the lake beds and some downward faulting of the land also took place. This created a basin which was the beginning of the formation of the Great Artesian Basin that now lies beneath the desert (pages 92–93).
- More downward folding of the land took place 225 million years ago with freshwater lakes forming in the low points of the basin. Rivers fed into these lakes and sediments were deposited on top of the older sediments. These sediments developed over time into impermeable rocks (i.e. rocks that will not let water soak into them or pass through them — they are a barrier to water movement).
- In the Jusassic period 140 million years ago, sand deposition took place in the basin above the impermeable rocks. These sands were to become the aquifer rocks (rocks where water is stored) of the Great Artesian Basin.
- 100 million years ago much of Australia was again flooded by the sea. Marine sediments were deposited on top of the Jurassic sands. These formed the impermeable cap rocks above the porous (i.e. allow water to move through them) Jurassic sands. Over a period of 170 million years, between 270 and 100 million years ago, the Great Artesian Basin was formed.

- The huge Lake Eyre Basin formed about 70 million years ago by earth movements led to down-warping (sagging) of the crust. Shallow seas and lakes covered the basin and were fed by rivers. Two types of sediments were deposited: alluvial and lacustrine. The alluvial sediments were deposited by rivers flooding across plains and lower valley areas; the lacustrine deposits were the fine sands and gravels deposited on the lake bed. The deposits were as much as 200 metres thick. These sediments now lie beneath the modern-day Simpson Desert sand plains and dune fields. These sediments later became the source of the fine sand that would be reworked by wind (erosion, transport and deposition) when the 'modern period' of aridity returned to the area.

- Beginning around 2 million years ago, big changes in climate took place: rainfall decreased and an arid climate developed. The lakes and rivers dried up.

- The modern Simpson Desert sands and dunefields are young in geological terms. Geological investigations suggest the dunefields began forming only 18,000 to 10,000 years ago. The sand covers the surface to a depth of between 1 and 10 metres. These sand plains and dunefields have a scattered cover of drought-resistant grasses and bushes, which bind the sand and prevents much movement of the sand from taking place (Figure 44). These sand plains and dunefields now dominate the desert surface landscape. After rains, the desert surface can be transformed by a cover of daises and other plants, which quickly die once the rains stop and wait to be regenerated in the next rain episode (Figure 45).

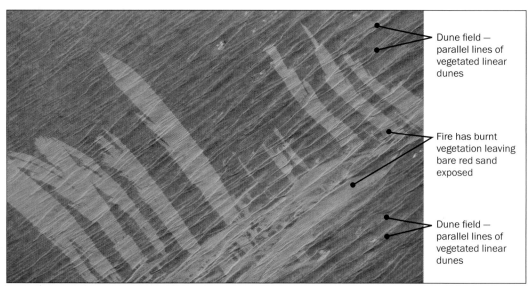

Dune field — parallel lines of vegetated linear dunes

Fire has burnt vegetation leaving bare red sand exposed

Dune field — parallel lines of vegetated linear dunes

Figure 44 Satellite view of the Simpson Desert linear dunes after fire.

Figure 45 Desert daises across the Simpson Desert after a storm.

 ISBN: 9780170233316

Case Study 3

Outback desert closed to protect tourists

November 2008

- **Three-and-a-half-month ban to deter crossings**
- **Summer temperatures forecast to reach 58°C**

Tourists will be banned from a vast area of the Australian outback for the first time this summer to prevent deaths from the extreme heat. The government announced that the Simpson Desert — situated in the dry, inhospitable heart of Australia — will be off limits from 1 December to 15 March.

In the height of summer the temperature of the desert sand reaches a blistering 95°C, capable of burning through shoes, said Joel Fleming, who has risked his life to rescue stranded tourists from the region.

'If you break down in that heat you're dead meat,' said Fleming, who has lived in the region for 52 years. 'You have nowhere to hide, no air-conditioning, no shade and no shelter.' Tourists, who need to drink 15 litres of water a day, can perish within hours, he said.

Known as the 'dead centre', the Simpson Desert is the driest part of Australia, with an average rainfall of 200 millimetres. It has no official roads, just tracks which criss-cross some of the world's longest sand dunes.

Mr Fleming said many European tourists tried to cross the Simpson Desert each summer but only about 40 made it. They did not understand the extreme remoteness, stifling heat or the tyranny of distance, and many were forced to turn back.

Learning Activities

1   Write a list of six sentences giving facts about the Simpson Desert. Each sentence must start with a word beginning with the letter 'S'. The first one has been completed as an example:

    i   **S**alty playa Lake Eyre is located on the edge of the Simpson Desert.

2   **a**   Why are there places in the Simpson Desert with an all-year-round supply of natural fresh water available?

    b   Why do colourful flowers sometimes cover large areas of the desert surface?

    c   Why is most of the desert surface a red colour?

    d   Why do the longitudinal dunes of the Simpson Desert run in a SSE– NNW direction?

    e   Why are the Simpson Desert sands and dunefields described as being 'young'?

    f   Why is the desert surface in the satellite picture (Figure 44) a mix of two different colours?

    g   Why has the Simpson Desert been closed, with people being told not to enter during summer months?

3   Make a copy of Figure 41 and then in a paragraph explain how longitudinal (linear) sand dunes are formed.

4   Use the dune photo (Figure 42) and Birdsville topographic map (Figure 43) to answer these questions.

    a   Describe the landform and 'water' features you would expect to see if you were on a visit to Birdsville and the surrounding area.

    b   Describe the location of Birdsville — use the maps in Figures 20 and 23 to give a long-distance (small scale) view and Figure 43 to give some (large scale) close-up detail.

History of the desert — how the Australian desert environment was formed and changed over time

Important understanding: The desert story is a two-part one: the present-day natural environment has resulted from the operation of 1. landform shaping (geomorphological) processes and 2. climate processes operating over millions of years.

Going back in time

The landforms of the present-day Australian desert environment are the result of a long history of landform development. The desert uplands, isolated rock outcrops, salt lakes, stony desert, sand plains and dune fields have developed over a long time period.

Although many of the landforms are the result of weathering and wind processes operating under present-day arid climate conditions, the influence of water and rivers operating in the past when the climate was wetter have also been important.

The past in the present

A The recent past and the present: in the last one million years, the desert area climate has been 'dry' but has varied from being very arid (drier than it is today) to less arid (wetter than it is today). During the wetter periods of higher rainfall, fluvial (water and river) processes have influenced the landforms, eroding in upland and river valley areas and depositing across plains and in basins and lakes. The erosion caused the wearing down of many of the upland surfaces so no high mountains exist across the desert environment today. The deposition provided a large amount of sand across the plains and basins that wind action would later rework into features like sand plains and dunes when the drier conditions returned.

   The wetter phases have been caused by changes in the location of global climatic belts that have shifted Australia away from the influence of the subtropical high-pressure systems (page 86). Drier times than the present day have come during global ice ages. The most recent of these was at its coldest 20,000 years ago and this resulted in super-dry conditions across the arid centre of Australia.

   All of the present-day landforms (such as dunes and salt lakes) associated with aridity are less than one million years old. Some are a result of processes operating during just the last 20,000 years.

Figure 46 Cracked earth on dry bottom of a salt lake near Lochiel, South Australia.

B Further back in time: between 55 and 10 million years ago, Australia was moving slowly across the surface of the earth (continental drift), and travelling towards its present global position and away from Antarctica to which it had once been joined. Parts of central Australia had arid climate conditions during some of these times but there were periods when climate was warmer and wetter than today. During the warm and wet periods, chemical rock weathering took place in ideal conditions. Deep iron-rich soils were formed during this time, and these were to form the deserts of the future. Since their formation, these soils have been leached of most of their minerals leaving them as low-quality desert soils of today.

C The distant past: 400 to 100 million years ago, central Australia went through times of being covered by the sea leading to sediment build-up on the sea floor which was later to become land sediments (sands and gravels) when sea levels fell again. Folding and faulting caused land movements upwards and downwards in the central desert area. These two processes also caused fracturing (breaking and cracking) of the rocks which in later times resulted in more speedy weathering and erosion. During this time period, landforms like Uluru and Kata Tjuta were pushed upwards by the land movements. After this, as the rocks around them were slowly eroded away, the resistant and hard rock areas were left exposed and upstanding above the eroded land around them.

Figure 47 Kata Tjuta.

In contrast to the many changes that have affected the central desert areas, the Western Plateau desert area (Figure 15), including the Great Sandy Desert and the Kimberley and Hamersley Ranges, has existed as a landmass for more than **500 million years**. The Western Plateau desert area is made up of rocks formed over 3500 years ago during the early years of planet earth's existence. Over time these rocks have been uplifted, folded, weathered and eroded into a red-coloured and mostly flat land, rich in the minerals that power Australia's mining industry but poor in plant nutrients.

The five case studies describe and explain how landforms and climate have changed over time in different locations within the desert environment:

- Uluru (pages 68–77)
- Triodia desert grasslands (page 90)
- The Great Artesian Basin (pages 92–93)
- The Simpson Desert (pages 98–101)
- Kati Thanda–Lake Eyre (pages 104–106).

Figure 48 Carr Boyd Range, Kimberley, Western Australia.

ISBN: 9780170233316

Kati Thanda-Lake Eyre

Case Study 4

The Aboriginal name for Lake Eyre is Kati Thanda and the lake now has the twin official name of Kati Thanda-Lake Eyre.

The map and photo of Lake Eyre (Figures 49 and 50) are deceptive. They give the impression of a large wet area located in central Australia. Nothing could be further from the truth. The lake and basin are located in the driest area of Australia where rain seldom falls and the lake is rarely full of water (Figure 51).

Lake Eyre is a non-permanent lake fed by non-permanent rivers located in the southwest of the Simpson Desert.

Lake Eyre is an example of a desert landform feature called a 'playa'. A playa is an area of flat salt-covered land in the centre of an inland drainage basin. When the lake fills with water, it is called a 'playa lake'.

The centre of Lake Eyre is the lowest point in Australia, 16 metres below sea level.

The drainage basin of Lake Eyre is a closed internal one — instead of flowing outwards towards the sea, the rivers flow inwards (inland) and empty into low areas in the centre of the basin and create lakes. There is no connection to the sea and coast at all. The Australian desert area has lots of these internal basins and lakes. Lake Eyre is the largest. The drainage basin of Lake Eyre covers one sixth of Australia but the rivers that feed Lake Eyre within this basin and the lake itself are usually dry. The river beds fill with water and flow towards the centre point of the basin filling the lake with water only in wet years and

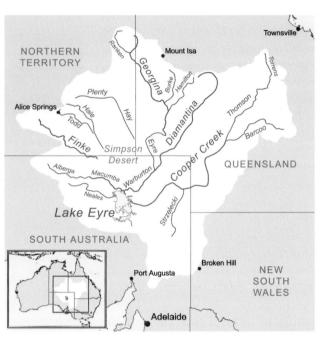

Figure 49 Lake Eyre drainage basin.

Figure 50 Lake Eyre filling after heavy rain.

Figure 51 Lake Eyre dry surface with salt crust.

 ISBN: 9780170233316

after periods of heavy rain. Lake Eyre has significant amounts of water in it only once every five to ten years and gets a complete fill only two or three times a century.

Because the lake waters are shallow and the temperatures are high, the process of evaporation aided by infiltration quickly remove water from the rivers and lake. Evaporation leaves behind minerals contained in the water and these build up to form a surface salt crust as much as 50 cm deep (Figure 52).

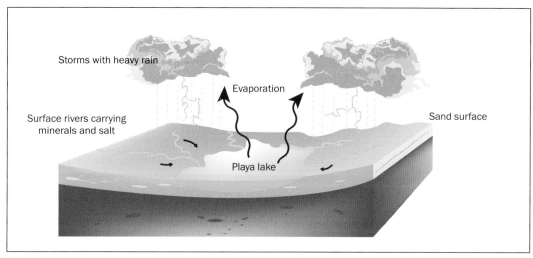

Figure 52 Playa lake formation.

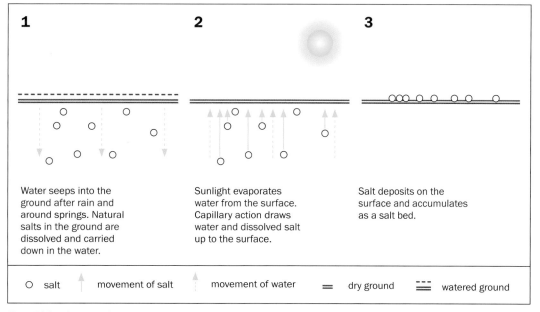

1 Water seeps into the ground after rain and around springs. Natural salts in the ground are dissolved and carried down in the water.

2 Sunlight evaporates water from the surface. Capillary action draws water and dissolved salt up to the surface.

3 Salt deposits on the surface and accumulates as a salt bed.

○ salt   ↑ movement of salt   ↑ movement of water   = dry ground   ≡ watered ground

Figure 53 Development of salt pans and saline surface soils.

How Lake Eyre has formed and changed over time

- 200 million years ago, land in the area where Lake Eyre was to form sank and became a large depression in the land surface. This low area filled with sediments deposited by rivers flowing into the depression.
- Between 100 and 25 million years ago, the area became covered by the sea as a result of rising sea levels.
- When the sea level fell and the area became land again 20 million years ago, large rivers flowed away from the Lake Eyre Basin area towards the south coast of Australia.

ISBN: 9780170233316

Case Study 4

- Around one million years ago, land movements faulted and tilted land to the south of Lake Eyre upwards. This new upland blocked the sea outlet of the south-flowing rivers. A huge lake began forming in the location of the present-day Lake Eyre. This new lake was many times larger than the present Lake Eyre and was called Lake Dieri.

- Between one million years ago and the present day, global climate went through a cycle of ice ages followed by warmer times. The climate of Australia was influenced by these changes in global temperature — it became colder and drier during the ice ages and warmer and wetter between the ice ages. Lake Eyre as it was to become known was influenced by these changes — during the wetter times it was a large permanent lake, during the drier times it was totally dry. During these drier times, lots of the surface sands of the lake bed were carried away by the wind and the lake bed surface became lower as a result.

- From around 20,000 years ago, the modern-day Lake Eyre developed, not permanently full of water or permanently dry. It is now a playa lake, an ephemeral (short lived) salt lake that varies in size from being completely dry some years to being a broad shallow lake in other years. In the dry years, wind continues to remove surface sediments but these are replaced by rivers bringing in and depositing sediments in the wetter years.

Learning Activities

1 **a** Read the paragraph starting 'Important understanding' on page 102. Which two processes have been important in the shaping of the Australian desert environment?

   b How have times of wetter past climate affected the present-day desert landforms?

   c Refer to 'Going back in time' and 'The past in the present' (pages 102–103). Complete this sequence of events flow diagram by selecting examples of events and features to put into each frame.

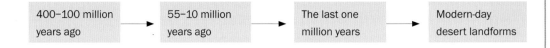

| 400–100 million years ago | → | 55–10 million years ago | → | The last one million years | → | Modern-day desert landforms |

2 **a** In what ways are the map (Figure 49) and photo (Figure 50) of Lake Eyre 'deceptive'?

   b Write a series of factual statements about Lake Eyre using these sentence beginnings:

   - *Lake Eyre is …*
   - *Lake Eyre has …*
   - *Lake Eyre can …*
   - *Lake Eyre was …*
   - *Lake Eyre may …*
   - *Lake Eyre will …*

   c Two events took place around one million years ago that had a big impact on Lake Eyre. What were they and how did they influence the lake?

   d Why is Lake Eyre today usually an area of dry salt pan (salt crust) rather than being full of water?

   e Make a copy of Figure 52. Using information from Figures 52 and 53, explain how playa lakes and salty dry lake beds form. The main thing is to explain where the surface salt deposits come from.

 ISBN: 9780170233316

People in the Australian desert environment

Important understanding:

- People perceive and use the Australian desert environment in many different ways.
- There is interaction between people and the environment.
- Perceptions and uses of the environment have changed over time.

The Australian desert environment, like all desert environments, presents difficulties for people because of the scarcity of critical resources like water, high-quality soils, food and animals, the temperature extremes and the limited amount of vegetation. The desert areas of Australia are also places where access is difficult — most of the desert is far from the coast and there are no rivers that can provide pathways into the interior. Travel across sand plains and sand dunes in searing temperatures with little surface water available is as challenging as climbing steep, snow-covered mountains. The desert environment of Australia is an inhospitable one for people.

In spite of these many challenges, the desert areas of Australia have a long history of occupation and use by people, and the way people perceive and use the environment is different now from what it was in the past.

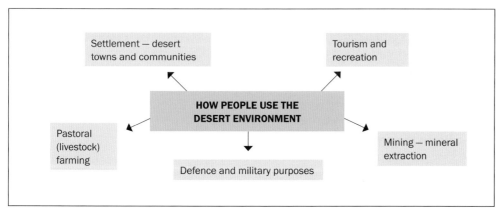

Figure 54 How people use the Australian desert environment.

Interaction between people and the Australian desert environment

People and the environment affect each other. Figure 55 shows the two-way nature of the interaction.

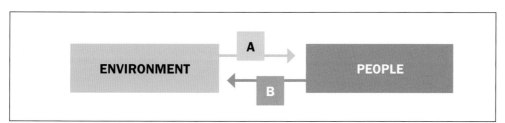

Figure 55 This diagram shows the environment affecting people (Arrow A) and people affecting the environment (Arrow B).

Each of the different ways that people use the desert environment provides an example of interaction taking place between people and the environment.

Example 1

Cultural interaction: Aboriginal people in the desert environment

The environment influenced the way people lived

Aboriginal people have lived in the desert area of Australia for around 40,000 years. They understood and cared for the environment and saw themselves as being part of the environment. They could find food and water even in the driest of times. Remoteness and accessibility was not a barrier to them. They could live and survive successfully in the arid environment. The people were desert experts.

They survived by following a nomadic hunting and gathering way of living within their own territory area. They moved to find food (plants and animals), leaving behind areas to regenerate the food supply. They moved to places with fresh water available either on or just below the desert surface. Traditional Aboriginal land use practices meant resources were used in such a way that they were renewed or preserved and not exhausted. This way of life continued for some families and communities

Figure 56 Traditional Aboriginal homeland.

into the 1950s. Since then the people have moved into towns and cities and to work on farms and in the mines. Some have become tourist guides and promoted Aboriginal arts and crafts centres. This land, their land, remains of great traditional and spiritual importance to them even though many have moved away.

Impact of people on the environment

The impact Aboriginal people had on the environment was limited. The main impact was through fire and burning. Fire was used for cleaning up and clearing vegetation. Without vegetation the land was easier to walk through, and safer too because snakes could be more easily seen and avoided. Fire was also commonly used to promote the growth of valued grasses and plants. The regrowth of grasses with fresh and soft leaves and foliage attracted grazing animals, such as kangaroos, back to the area for easier hunting. The use of fire to drive out animals which could then be killed for food was common practice, and smoke from fires was sometimes used to force possums from their hiding places in hollow trees.

Figure 57 Grasstree plant.

This controlled use of fire has been used for tens of thousands of years in the desert zone. This burning changed the appearance of the Australian bush — large areas of forest were replaced by open grasslands. Plant species which did not regrow well after fire declined in number and sometimes died out. More fire-resistant species, such as grasstrees (Figure 57), eucalypts and acacias became dominant forms of vegetation across many semi-arid areas.

 ISBN: 9780170233316

Learning Activities

1 **a** Read pages 107–108 and find the words in the text listed below in Column A. Copy the table but match the definitions in column B with the words in column A.

Column A words	Column B definitions
Scarcity	To try and to make an attempt
Critical	To look after and preserve
Access	Having a shortage of something
Searing	Very hot and burning
Inhospitable	Unpleasant and unwelcoming
Endeavoured	To be on the move from place to place
Cultivate	How easy it is to get to a place
Nomadic	Essential and most important

b Write a short paragraph titled 'People and the desert environment' that includes each of the eight words from column A in the table.

2 Why would Aboriginal people be described as 'desert experts'?

3 Make a copy of this interaction diagram but replace boxes A and B with examples from the way Aboriginal people traditionally lived in and used the environment.

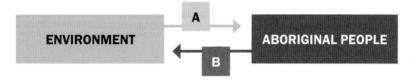

Example 2

Economic interaction: Desert living — communities and towns of the desert

The desert environment of Australia covers 70 percent of the land area of the country but has a very low population density. Out of a total population of 23 million, only 600,000 live in this area. The 'largest' towns are small — Kalgoorlie has a population of 30,000 and Mt Isa 27,000. Both of these towns are semi-arid mining towns located towards the edge of the desert area. Alice Springs, located near the centre of the desert environment, has a population of 25,000 and is the largest 'true desert town' located in an arid rather than semi-arid area (Figure 58). There are also some small communities of fewer than 1000 people and farm settlements scattered across the desert area but large parts are uninhabited.

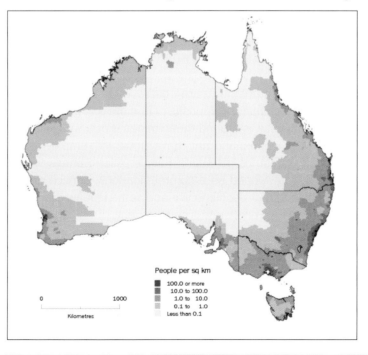

Figure 58 Australia population density and distribution.

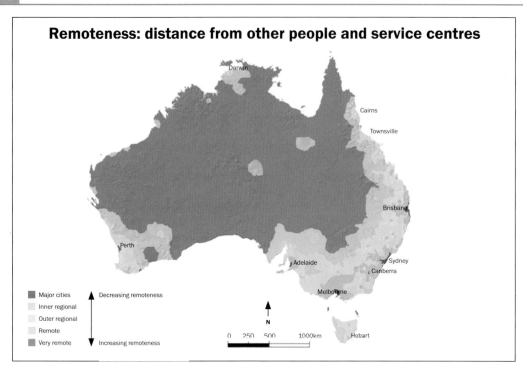

Figure 59 *The desert area is in the most remote (purple shaded) zone.*

The absence of people and overall low population density are an example of the environment affecting people. Environmental conditions make farming difficult, and in many places impossible. This plus a lack of employment opportunities and remoteness of the area help explain why so few people live in the desert environment.

The low population density and remoteness of the central desert area from large population centres has seen parts of the area used for weapons and nuclear tests in the past, and today for missile and rocket tests and launches and for military training (Figure 60).

Figure 60 *Entrance to the Woomera Test Range west of Lake Eyre in South Australia.*

The larger mining settlements are located where they are because of the natural environment (geology). The rich mineral deposits of iron ore and gold have made it economically possible to develop mines. It is around these mines that towns like Kalgoorlie (gold mining) and Mt Isa (lead, silver, copper and zinc) have grown with employment opportunities in both mining and service industries attracting people. Alice Springs with its 'middle of the continent location' has developed as a communications, service and tourism centre. Employment opportunities attract people to live and settle in Alice Springs even though it is located in a remote area with a challenging environment.

The influence of people on the environment can be seen in the 'townscapes and minescapes', which when viewed close up and from above are totally different from the surrounding natural desert landscape (Figure 61).

In summary, the environment has influenced population density and settlement location, and in turn people have impacted on the landscape with town and mine developments.

Figure 61 Super Pit at Kalgoorlie, the largest open-pit gold mine in Australia.

Learning Activities

1 a Refer to Figure 58. Describe Australia's population distribution pattern.

 b i Refer to Figure 59. How does 'remoteness' vary across Australia?

 ii How has remoteness been measured to create this map?

 c Either describe the scene shown in Figure 61 (Super Pit) or draw an annotated sketch of this photo.

2 Copy the 'In summary ...' sentence at the bottom of page 110. Give some facts and examples to support this summary.

Example 3

Economic interaction: Farming — natural environment links

Farming takes place across a large part of the desert area, especially the semi-arid areas. Not any type of farming though. The natural environment has influenced the type of farming — farming is adapted to the natural environment conditions. Huge cattle and sheep farms called stations are the 'farming type' of the desert area. The largest stations are as big as small countries and similar in size to provinces in New Zealand like Northland and Marlborough. The farms and farming system are a response to the environment — aridity means water is scarce and vegetation sparse so that a large amount of country is needed to support enough cattle and sheep to make a living. The number of livestock carried per hectare of land is very small, as low as one head of cattle on 100 hectares of land in places.

Figure 62 Mustering on an Australian station.

Case Study 5

Anna Creek Station

Anna Creek station located just west of Lake Eyre in South Australia is 24,000 sq km in area, a similar size to the whole Waikato region. It is the largest station in Australia. The stocking rate is very low — the huge station carries just 10,000 head of cattle in a typical year. The breed of cattle on Anna Creek is Santa Gertrudis (Figure 63). This breed was introduced into Australia from the USA. and is able to cope with harsh conditions, especially heat and drought.

The 'grass' on Anna Creek is the natural vegetation and unimproved, and cattle are usually born and grow up with little human contact. Because of the huge size of the stations and challenging environment, mustering of the stock is increasingly done by using helicopters, light planes and trail bikes rather than the traditional horse. The rounding up of stock is essential to the operation of the station. Their sale provides the farm income. Once the cattle are mustered into the yards, vet inspections and branding take place. Stock to be sold get separated and loaded onto road train trucks (Figure 64) to be taken to saleyards in larger centres.

The rest return to their 'million-hectare paddocks'. Stock numbers on the stations vary greatly from year to year. The farm managers destock and sell the cattle during the bad drought times, and restock only when more water and feed become available after wetter years. Since the year 2000, cattle numbers on Anna Creek, for example, have been as low as 1500 during the bad droughts of 2006–2008 but as high as 17,000 during the wetter 2012–2013.

Figure 63 Santa Gertrudis cattle.

Anna Creek has its own small airstrip and this plus the use of satellite technology reduces the isolation of the station. It also allows easy access for the Royal Flying Doctor Service if a medical emergency occurs.

Figure 64 Road train transporting cattle.

For children living on the station, their school is 'School of the Air' using radio, computer and satellite technology. The centre of the station is the 'homestead', where the manager lives. Surrounding the homestead are the machinery sheds and accommodation for workers plus the stock yards.

A

B

Figure 65a and b Anna Creek homestead, accommodation and machinery shed.

   ISBN: 9780170233316

People have impacted on the desert environment through their farming practices and activities in ways often not intended by the farmers. Land degradation is one such example. Land degradation means the reduction or loss of the biological or economic productivity of the land due to the actions of people. An extreme form of degradation is desertification where the land ends up desert-like and useless. Land degradation has occurred in as much as 50 percent of the semi-arid area where pastoral farming has been developed. Figure 66 shows causes and results of the degradation.

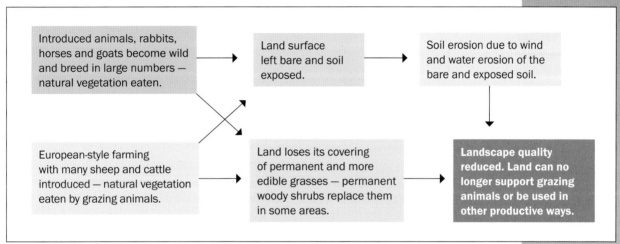

Figure 66 Land degradation.

Another feature of land degradation is that it is not even across the desert environment — there is a spatial pattern (Figure 67). Most of the sheep and cattle when first introduced were concentrated in places near to natural water supplies from water holes and springs, and later in places where bore water was available. In these 'wetter' areas there is both a more reliable supply of drinking water and more natural grass available. These locations have in the past suffered the greatest amount of land degradation. They were overstocked — there were too many cattle or sheep for the water and grass amounts available during times of severe drought. When long droughts occurred, firstly all the vegetation was eaten to the point where any chance of regeneration when wetter periods returned was lost. Secondly, huge stock losses (deaths) took place. Today, with improved roads and improvements in desert area transport, stock can be removed from drought-affected areas, which benefits both the stock and vegetation.

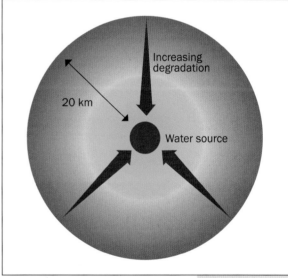

Figure 67 Land degradation.

Learning Activities

1   List five distinctive features of Anna Creek Station.

2   Explain how the operation of Anna Creek (the way the farm is run) is influenced by its being located in a desert environment.

3   a   Make a copy of Figure 67.

    b   Write a one-sentence definition of the term 'land degradation'.

    c   Explain why land degradation often occurs in areas next to and close to water sources.

People and the Great Artesian Basin (see also pages 92–93.)

People interact with the Great Artesian Basin (GAB) in ways that extend well beyond farming.

Water flowing to the surface from the huge natural underground reservoir has been vital to the development and opening up of the desert area. The underground water flowing naturally to the surface at springs and at water holes has been added to by the sinking of bores to tap and pump to the surface even more water. People rely on this water.

- The water was and remains vital to Aboriginal desert communities. They continue to maintain a connection with water and have a holistic approach to the environment that is difficult for people coming from a Western perspective to understand. The Dalhousie Springs fed from the GAB are part of the traditional lands of the Arrente people. The springs in Arrente language are known as 'irrwanyere' or 'the healing springs'. The people value the water for its healing powers. For the Arrente, the springs are not just a source of drinking water and fish, but also a travel path that goes deep underground which gives them spiritual and cultural links with their lands and connects them with other indigenous groups who live in the GAB area. An Arrente elders says: 'We are in the middle of kwatyc (water), it is all around us, we have to look after this place.'
- More recently the water allowed the development of the cattle- and sheep-grazing industry. Pipes and tanks on the stations, fed from the natural underground reservoir, supply the stock as well as the homesteads with their water. Today many small towns and communities, mines and tourist operations are dependent on this water.

Figure 68 Artesian water bore and pumping station at Thargomindah, South West Queensland.

The first artesian bore of the GAB was sunk in 1878. Now there are over 4500. European settlers thought the supplies of artesian water were inexhaustible. So many wells were drilled and allowed to flow freely that water pressures dropped. Many bores and springs stopped flowing. With open wells and open drains, 90 percent of the water coming to the surface is lost through evaporation. This way of using the resource was not sustainable (Figure 68). Artesian wells and bores that once flowed at rates of over 10 megalitres per day now provide only between 0.01 and 6 megalitres per day. Pressure and flow rates have decreased all across the basin (Figure 69). About one third of the wells that were artesian (flowing under natural pressure) when drilled have now ceased to flow and require pumping.

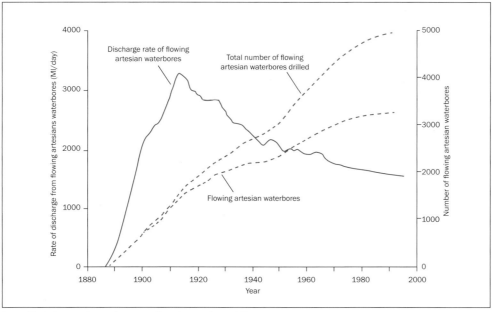

Figure 69 Underground to surface water flow changes in the GAB.

Learning Activities

1   Explain why water from the GAB is so important to life and people in the desert area.

2   **a**   Describe and explain the different trends shown on the graph in Figure 69.

b   What future problem is likely to occur based on the trends shown?

c   Suggest how this future problem could best be planned for, or resolved, before it becomes a major issue and cause of ongoing concern.

Perception of the environment — how people view the environment and how views have changed over time

Perception is about how people see things and the views and opinions people have about things. In geography, perception is to do with the way people view and interpret places and the environment and how they think about and view geographic issues.

Perception is often influenced by perspective — who we are and the position we come from influences the way we view places, the environment and the opinions we have about geographic issues.

Important understanding: In the past, the desert environment was perceived by European explorers and settlers as being inhospitable, dangerous and an area that was hard to access. It was a land to be avoided and seen as offering little opportunity for work or wealth. All this has changed.

Changes have involved the following:
- Huge cattle and sheep stations were developed and found to be profitable.
- Vast and valuable mineral deposits have been discovered and mined. Much of Australia's wealth and growth relies on the mineral resources of the desert.
- Access has improved — road, rail and air travel have opened up the area. The Internet and satellite communication have lessened the remoteness of the area.
- Tourism has become an important and growing desert industry. The aridity and remoteness that once kept people away from the area are now the things that attract the 21st-century tourist. People still come for the stunning scenery and to experience the unique environment, but increasing numbers want to experience the open spaces,

quietness and solitude away from the crowds of the coastal cities — they want something different. A totally new type of tourism is also developing, catering for the those who want something extreme, and to challenge themselves against the environment.

Tourists and Australian deserts

Some features of Australia's deserts are well known and visited by many tourists like Uluru and Alice Springs. The number of tourists who visit the less well-known desert areas has been small, but is growing.

Attractions of deserts

The main tourist attractions in Australian deserts are the spectacular sand and rock features. They are illustrated in tourist brochures and people travel long distances to see them. Increasing numbers of people want to visit desert country that is less well known like the Simpson Desert and Lake Eyre Basin.

Tourist needs

Facilities need to be built to cater for the influx of tourists visiting the deserts' attractions. Desert areas in Australia have few large settlements. Therefore the accommodation, shops, food supplies and transport provisions that are needed by tourists have to be specially established.

Ecotourism

More people want to visit the desert without being in the company of hundreds of others. They do not want luxurious accommodation and entertainment. Such tourists want to see more remote areas of the desert and become involved with indigenous culture. These are characteristics of ecotourism, a form of tourism favoured by people who want to be in natural surroundings without harming the environment. One of the main resources required by ecotourism is skilled operators with a high level of knowledge of the area and an understanding of the ways to conserve its environment. Aboriginal knowledge and expertise is being used in the development of a growing number of indigenous tourist ventures.

Figure 70 Tourism in the desert environment — a fast-growing industry.

Case Study 6

Kilcowera Station near Thargomindah, South West Queensland

Working cattle place and farm stay

Imagine sitting by your campfire as a vivid, red sun sets over a timeless outback landscape. Parrots chatter in nearby coolabah trees as the purple dusk creeps across a cloudless sky. This is Kilcowera Station — a stunning contrast of mulga rangelands and ephemeral wetlands brimming with birdlife, in South West Queensland. Halfway along the Dowling Track, Kilcowera is adjacent to the Currawinya National Park and offers exceptional views and access to Lake Wyara.

Case Study 6

While this is truly a birdwatcher's paradise, other animals to be seen include red kangaroos and eastern greys, the western grey kangaroo, wallaroos, emus, lizards, echidnas, yabbies and small carnivorous marsupials.

Figure 71 Camping at Kilcowera Station.

A comment from the station owners (2011): *Our little tourism business is steadily growing; we feel we must be doing something right as we have repeat visitors who appreciate the improvements we are constantly implementing around the place. We do not aspire to be too commercial or expensive but want to retain the outback feel and continue to offer guests our time, hospitality and knowledge of our place and the local area ... The countryside still looks fantastic after the rain earlier this year, let's hope it keeps on raining! Please, please! We would like a few good seasons rather than more drought. We are trying to buy some cows to restock the property but they are so expensive, we can't justify spending a thousand dollars for a cow and calf.*

Testimonial: 'A wonderfully presented destination, the information/interpretation is great. We especially enjoyed our self-drive tour of Kilcowera and the fabulous camp oven meal. Thank you for your hospitality, warm welcome and for sharing your magnificent property with us.'

Figure 72 Wildlife at Kilcowera Station.

Big Red Run

When distances being run are marathon plus, day in, day out; when the territory is both as brutal and other-worldly beautiful as the Simpson Desert; when the forces that attempt to halt your progress attack from within your body — fatigue, blisters, torn tendons, bruised feet — and from without — heat, sand, wind and no water — you need help. So it was a fitting end in Birdsville, Queensland, to the first

Figure 73 Extreme and adventure tourism.

(2013) 250 km Big Red Run multiday with the entire field running as a closeknit group down the broad, dusty main street of Birdsville to finish on the steps of the iconic Birdsville Pub. They ran into the bar and a few cold beers to boot as a newly formed family group, each having conquered the same overall obstacles of the desert along with their own, very personal demons of mind and body to finish an adventure run odyssey like no other.

Course description: The eastern Simpson Desert is a flat landscape with stable sand ridges laid over the top. Most of the dunes are firm and wind-packed with hard sand and gibber plains in the valleys. The dunes are covered in light vegetation — sparse grasses, spinifex and bushes that bind the sand. The tops of most dunes have loose, wind-blown sand.

Figure 74 Start of the Big Red Run in Birdsville.

More than 95 percent of the course is in areas of the desert away from vehicle tracks, trails or signs of humans. Unique among off-road running events, you'll be following a specially marked route cross-country in unchartered and untracked desert. This adds to the adventure of the Big Red Run.

Event	Surface and description	Suitable for ...
Big Red Dash 42.2 km marathon	Approx. 70% gibber plains, 20% firm surface sand (flat valleys and dunes), 10% loose sand on dune crests.	Runners, trekkers, novices. The mostly firm surface, beautiful surrounds and generous cut-off time make it suitable for all walkers and runners.
Born to Run 100 100 km ultra	Approx. 20% gibber plains, 70% firm surface sand (flat valleys and dunes), 10% loose sand on dune crests.	Very fit runners and trekkers. The cut-off times will allow a fit walker to complete the entire course in a long day.
Big Red Run 250 km 6-day stage race	Approx. 15% gibber plains, 75% firm surface sand (flat valleys and dunes), 5% loose sand on dune crests and 5% following the dry bed of Eyre Creek.	Very fit runners and trekkers. Daily cut-off times will allow a fit walker to complete each day's stage.
All three events also include salt lakes and flat clay pans.		

Figure 75

A new approach to sustainable desert land management

Viewing the natural environment through a different lens/from a different perspective

The environment has not changed but the way people perceive it has. Severe and ongoing drought was once thought of as both catastrophic and 'unlucky', an event that could not be predicted or planned for. Affected farmers received financial help when drought was declared. This approach did little to encourage farming methods that would be sustainable for both the farmer and the environment. Old farming practices continued.

Figure 76 Scientific research and a new approach to desert management.

A new approach based on scientific research has become government policy — to recognise drought as an accepted part of farming in the semi-arid grazing areas, something to be expected and something that can be planned for. Cattle and sheep farmers have been encouraged to adopt self-reliant approaches to manage for climatic variability, and to maintain and protect Australia's agricultural and environmental resources during periods of extreme climatic stress. This approach has focused property managers on adopting risk management strategies to reduce the impact of drought on both agricultural production and the environment. Financial help is made available only when conditions became extreme and beyond what could reasonably be expected in an environment known to be dry and unpredictable.

Graziers have been given access to a range of crop, grassland and pasture models and maps developed by official agencies based on field research, ground monitoring stations and satellite remote sensing data to assist in tracking agricultural conditions such as pasture cover, nutrient availability and meteorological conditions.

This information is designed to encourage station managers to view the environment differently and to employ risk management techniques to reduce the environmental stress caused by long drought events. One action that can be taken, for example, is to reduce stock numbers before the drought becomes severe. This reduces the long-term negative impacts on both the stock and environment. Planning for the predictable has become the new focus.

 ISBN: 9780170233316

Aboriginal stewardship

Over the past 150 years, Aboriginal communities have lost ownership and control of their desert lands. Developments in the desert area (pastoral farming, mining, road construction, military uses) took place as if 'nobody' owned the land and as if the land was 'empty' when Europeans settled. Now there is a new perspective, one that acknowledges Aboriginal people as the original people of the land and the rightful owners. This has meant two things:

a Aboriginal environmental knowledge and land use practices have been recognised and valued as showing ways to use and live in the desert environment in a sustainable way.

b Stewardship and co-management of many land areas and cultural sites has been given back to Aboriginal communities. Tourism ventures have been developed run by Aboriginal people to share their desert knowledge, culture and food with tourists from across Australia and from overseas. A feature of this new approach is the establishment of two-way links between indigenous land managers and government agencies involved in sustainable land management and conservation of the desert area. Indigenous Land Management Facilitators give support and research advice to Aboriginal communities about management of their land and cultural sites. The facilitators in turn provide information and feedback to government authorities about issues of concern and visions Aboriginal communities and managers have about current and future use of their lands and cultural sites.

Figure 77 Aboriginal run tourism at Uluru.

An Aboriginal elder summed up the view they have of the land:
'We cultivated our land, but in a way different from the white man. We endeavoured to live with the land; they seemed to live off it. I was taught to preserve, never to destroy.'

Learning Activities

1 In the past, many Europeans had a negative view of the desert area, but now the view is more positive. Why has this change in perception come about?

2 Refer to the Kilcowera study on pages 116–117:

   a What big change has taken place on the station?

   b What type of tourist would most likely want to visit the station? Give reasons for your answer.

   c Draw a star diagram showing four selling points (attractions) of Kilcowera.

   d What evidence is there that cattle grazing remains an important part of the station operation?

3 Refer to Big Red Run on pages 117–118. Describe the course you would run if you entered the six-day event.

4 Choose either 'A new approach to sustainable desert land management' or 'Aboriginal stewardship'. Describe what the new perspective is and how this is bringing about changes to the way people use and manage the desert environment.

5 **a** Copy the 'Important understanding' from page 115.

   b Give an example of how this 'old perception' influenced the way people used or approached the environment in the past.

   c Write a statement about how people view and perceive the environment today. Give an example of how this has influenced the way people approach and use the environment in a way that is different from the past.

The Detroit of today: the decaying Detroit and the prosperous Detroit.

Understanding Motor City –

Urban Patterns: Detroit, USA

3 Motor city Detroit, USA

Figure 1(a–d) a 1919 Model T Ford, b 2014 Chevrolet Camaro, c modern-day downtown Detroit, d suburban Detroit.

 ISBN: 9780170233316

Focus on a temporal pattern and three centuries of population change

Detroit — a puzzle

Charlie LeDuff wrote in the preface to his book *Detroit: an American autopsy* (2012), 'Having led us on the way up, Detroit now seems to be leading us on the way down. Once the richest city in America, Detroit is now the nation's poorest.'

A British reporter was amazed by what he saw on a recent visit to Detroit. He wrote: 'Much of Detroit is dangerous for its own residents, who in many cases only stay because they have nowhere else to go. The isolated, peeling homes, the flooded roads, the clunky, rusted old cars and the neglected front yards amid trees and waste-high grassland make you think you are in rural Alabama, not in one of the greatest industrial cities that ever existed.'

What has caused the decline of this once great city?

Locals in Detroit today describe the city as 'like living in a museum — a museum of neglect'. While cities across the world plan for population growth and creating larger and better cities, Detroit faces different challenges. It is a city that had a dramatic rise and then a spectacular fall.

In less than 60 years the population of the city has dropped from a high of nearly two million people to a city today of little more than 700,000 people. It is a city that has 66,000 vacant sections and 78,000 abandoned buildings. It is a city with empty skyscrapers, factories left in ruins, and residential streets once densely populated but now full of houses left abandoned, burnt to the ground or demolished spread across its urban landscape.

Figure 2 Abandoned St Agnes church.

In 2009, the Detroit mayor's office introduced a policy called 'Transforming Detroit'. The policy set out plans for the redevelopment of the city, improving the quality of life of all citizens, repopulating the city and reclaiming the future for Detroit as a world-class city. The policy highlighted the need to:

- improve and restore the three essential services of public safety, public transportation and public lighting
- provide recreational opportunities for the people
- deal with urban blight and decay
- sort out the city debts and stabilise the city finances
- reorganise government and rethink urban policy.

This policy and the plans are a sign that Detroit is a city facing an unusual set of problems and challenges.

ISBN: 9780170233316

1902

2012

Figure 3 Detroit waterfront 1902 and 2012 — urban change over the last century.

Learning Activities

1 a In which US state is Detroit located?

b How many people lived in Detroit at its population peak and how many live there today?

c What things about Detroit amazed the British reporter?

d Copy the sentence that ends with the words 'urban landscape'.

2 a Write four words to describe each of the two Detroit landscape photos in Figures 1c and 1d. For example, downtown Detroit photo = riverside, high rise, ??; suburban Detriot photo = wasteland, abandoned building ??.

b What similarities and what differences are shown in the 1902 and 2012 photographs of Detroit in Figure 3?

3 Draw a star diagram to show the five aims of the policy 'Transforming Detroit'.

1:

5:

2:

Transforming
Detroit

4:

3:

4 Detroit faces unusual problems and challenges. Write what these are in fewer than 25 words.

5 What is the 'puzzle' referred to in the section 'Detroit — a puzzle'?

Background — famous cities

Famous cities of the world

Creating lists of the most famous and well-known cities is popular. Often the list is based on the opinion and knowledge of people in a survey. Cities like New York, Los Angeles, London, Rome, Paris, Tokyo, Beijing, Hong Kong, Sydney and Rio de Janeiro are common in such lists. More objective lists are produced from the use of official statistics. Ranking cities according to the number of tourists who visit or by the population size of the city are two ways of producing such lists (Figure 6). In tourist visitor numbers, Bangkok, London, Paris, Singapore and New York are the top five global cities. In population total, Tokyo, Seoul, Mexico City, New York and Mumbai make up the top five.

Cities often appear on top lists for having distinctive features connected with history, architecture or news reports. Athens with remains of Ancient Greek buildings, Venice with canals, Washington DC as the capital of the USA with the White House, and Dubai with stunning modern architecture are famous because of these distinctive features.

Detroit is famous for a different set of reasons

In the mid-20th century, Detroit was described as the 'city of the future'. It was a prosperous and progressive city that had been through a century of rapid growth. Detroit was the home of the 'big three' US auto makers Ford, General Motors and Chrysler. Being top in America meant these were the top three global auto makers too. These giant Detroit-based companies were famous all over the world. Detroit got nicknames of 'Motor City' and the 'Automotive Capital of the World'. The city also gained fame as the birthplace of Motown Records and Motown music.

Detroit's fame has remained since the 1950s but for all the wrong reasons — it has become a city in decline and freefall, a city that has suffered huge population losses and a city with enormous social and economic problems (Figure 5a and 5b).

Figure 4 Company logos made famous by Detroit.

Figure 5 a and b The modern Packard Plant auto factory of the past, now shut down and abandoned.

Millionaire cities — the world's largest cities

Urban growth (cities getting bigger in size) and urbanisation (an increasing percentage of people living in urban areas) have been features of global population in the last two centuries.

Ancient Rome around 2000 years ago is thought to have been the first city to have more than one million inhabitants. When the power of the Roman Empire declined, so did the population of the city. It was another 1500 years before any city again reached a population of more than one million. During the early 1800s, both Beijing and London did so and they became the first millionaire cities of the modern era.

In 1800, only 3 percent of the world's population lived in cities. Since then things have changed rapidly. By 1900, 12 cities were recorded with populations of more than a million. By 1950, the figure was 83. In 2014, over 400 cities have millionaire status and more than 50 percent of the global population live in cities. Cities with over one million people are found in every continent and most countries have at least one city of millionaire size. In Oceania, Sydney is the largest millionaire city, with a population of more than four and a half million. Auckland is the only New Zealand city on the million plus list.

Key terms

City A large and mostly built-up area (large town) that is controlled and managed by one council or one local authority. The Auckland 'super city' is one example of such a city. In some places like Great Britain, a large town with a cathedral is called a city. In other countries, a population number, such as 50,000 or 100,000, distinguishes a town from a city. In New Zealand, towns with a population of more than 20,000 people were called cities, but this rule is no longer used in an official way.

Urban area A large built-up region that consists of a city and the suburbs and smaller towns that are around it. The Wellington urban area, for example, is made up of Wellington City plus Porirua, the Hutt Valley and Kapiti Coast.

Conurbation A very large built-up area where growing cities have joined together to form one large continuous built-up area. In Figure 6, the New York-Philadelphia area in the USA, the Rhein-Ruhr area in Germany, and the Osaka-Kobe-Kyoto area in Japan are examples of conurbations.

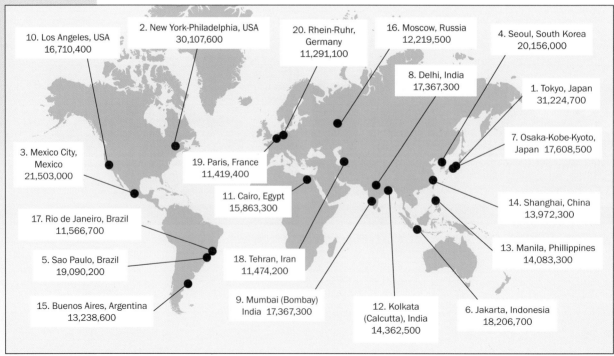

Figure 6 The world's 20 largest urban areas.

Learning Activities

1 **a** There are many lists of the world's 'top five' or 'top 10' cities. How are such lists created?

b In the section on 'Famous cities of the world', which city is named most frequently in the different lists?

c What does the term 'millionaire city' mean?

d Which continent has the greatest number of the world's largest top 20 urban areas located within it?

e Refer to the text and Figure 6 to help you complete the table.

Region/ Continent	Largest urban area	World ranking by population size	Country	Population size of this urban area in millions
North America	New York- Philadelphia		United States	30.1
Central America		3		
South America				19.1
Europe			Russia	
Africa				
Middle East	Tehran			
Asia		1		
Oceania		Not in top 20		

2 Draw three diagrams to show the key features of cities, urban areas and conurbations. (These diagrams will be like 'visual dictionary definitions'.)

3 Follow the arrows to unscramble this text to reveal three things that have made Detroit a famous city. Punctuate your unscrambled text and then give your text a title.

O	F	S	U	O	M	A	F	S	I	T	I	O	R	T	E	D	←	START
→	R	B	E	I	N	G	A	C	A	R	M	A	K	I	N	G	C	
O	M	E	R	E	H	W	Y	T	I	C	E	H	T	Y	T	I	←	
→	T	O	W	N	M	U	S	I	C	B	E	G	A	N	A	N	D	
E	N	I	L	C	E	D	N	I	Y	T	I	C	A	W	O	N	←	
→	W	I	T	H	M	A	N	Y	P	R	O	B	L	E	M	S	•	

Detroit city and its location

Important understanding: In 1950, Detroit was one of the 83 cities in the world with over one million residents. Its population total was similar in size to Sydney and three times the size of Auckland. Today, Detroit would not even make the list of the largest 500 global cities. Sydney now is seven times larger than Detroit and Auckland has double the population of Detroit.

Figure 7 The flag of Detroit.

The flag of Detroit makes links to the French and British origins of the city. The bottom left quarter with five gold fleurs-de-lis on a white background represents France and the top right quarter with three gold lions on the red background represents Britain. The top left and bottom right sections of the flag have 13 white stars and 13 red and white stripes representing the founding 13 colonies (states) of the United States.

The centre circular seal on the flag is in recognition of the great fire that destroyed the city in 1805. The person on the left is crying over the fire-damaged city; the person on the right is pointing to the new city that will rise in the future. The two Latin mottos mean 'We hope for better things' (*Speramus meliora*) and 'It will rise from the ashes (*Resurget cineribus*). Given the present-day problems the city faces, these two mottos seem to be apt and relevant to the Detroit of today.

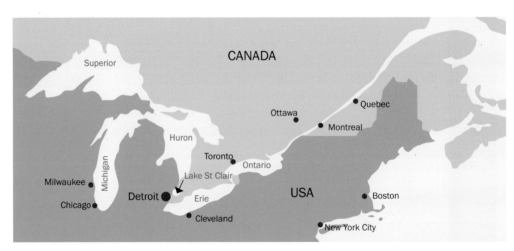

Figure 8 Detroit, an inland Great Lakes city.

Detroit is a city in the Great Lakes area in the northern part of the Midwest region of the USA. It is part of Wayne County and is the largest city in the state of Michigan. The city is located on the Detroit River, which links two of the Great Lakes, Huron and Erie. Although Detroit is located over 1500 kilometres inland, it is an important seaport. It has direct access to the Atlantic Ocean via Lake Erie, Lake Ontario, canals and the St Lawrence River. The city is close to the US-Canada border. The Detroit River forms the border between the USA and Canada. The Canadian city of Windsor (Ontario) is only a few minutes' drive away by bridge or tunnel across the river from Detroit.

 ISBN: 9780170233316

Figure 9 Detroit and surrounding area location features.

Learning Activities

1   **a**   Write a sentence that compares the population total of Detroit with the population total of Auckland between 1950 and the present day.

    b   How does the flag of Detroit reflect the history of the city?

    c   In what way are the two Latin mottos on the seal of the city relevant to present-day Detroit?

2   Using both the text and map information, either write a paragraph or draw an annotated map to highlight location features of the city of Detroit. Include both natural and cultural information in the answer.

Detroit's population over time: identifying patterns

Geography makes studies of two types of patterns:

Spatial patterns are about the way features on the surface of the earth are located, arranged and distributed. Spatial is to do with the 'space' of the surface of the earth. The way features are distributed on the surface of the earth usually fall into one of three categories: regular, random or clustered. Geography involves identifying spatial patterns and explaining them.

Temporal patterns are about the way things change over a period of time. In Geography the things studied are those that are connected with the surface of the earth, for example temperature change over time, land use change over time or population change over time. Temporal changes can be regular or irregular, fast or slow. Geography aims to identify the changes and explain them.

Graphs showing examples of temporal patterns:
how things change over a period of time

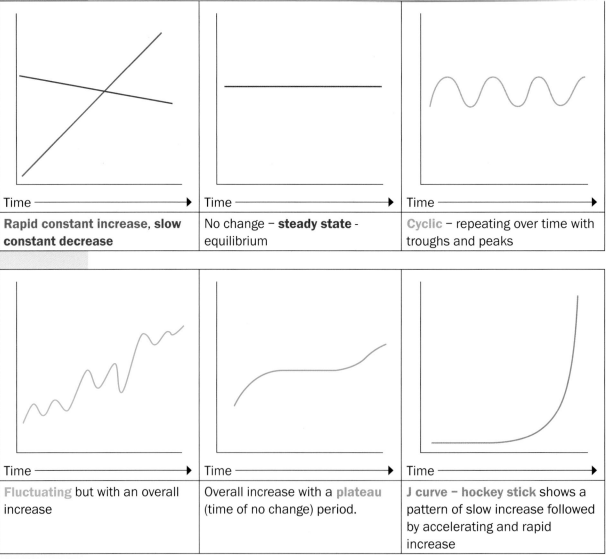

Time ————————▶	Time ————————▶	Time ————————▶
Rapid constant increase, slow constant decrease	No change – **steady state** - equilibrium	Cyclic – repeating over time with troughs and peaks
Time ————————▶	Time ————————▶	Time ————————▶
Fluctuating but with an overall increase	Overall increase with a plateau (time of no change) period.	**J curve – hockey stick** shows a pattern of slow increase followed by accelerating and rapid increase

Figure 10

Detroit's temporal pattern

The pattern being investigated in this study of Detroit is a temporal one: an investigation of how the number of people living in Detroit (the population total of the city) has changed over time.

Important understanding: Over the past 300 years there has been a pattern of growth in the population of Detroit. There was continual growth between 1700 and 1950, with the growth especially rapid between 1850 and 1950.

A variation in the pattern has taken place since 1950: population growth has reversed and a rapid and constant decline in the population total has become the new normal for the Detroit.

Detroit has an unusual feature in its pattern of population 'growth'

Like most cities, Detroit has more people living in it today than it did in the past. Since the city was first founded just over 300 years ago, the population of the city has grown. In fact, more people live in the city today than lived in the city 100 years ago.

A growing population from the past to the present is the basic and main pattern. This basic pattern can be divided into three phases (stages):

1 slow growth between 1701 and 1850
2 a speed-up in growth between 1850 and 1900
3 accelerating and very rapid growth between 1900 and 1950.

Figure 11 New cars built in Detroit loaded for rail transport, 1973.

After 1950, things changed and a complete variation to the pattern has taken place.

- Detroit has gone through a period of population decline — not just decline but a rapid decline.
- It is this decline that makes Detroit unusual, as most cities in the world have growing populations and in many cities the speed of growth is very fast (Figure 17).
- There are other cities that have lost population but few have lost as many people as Detroit.
- Detroit is the only city in the US to have exceeded a population of a million people only to reduce back again below that one million number.

Figure 12 Detroit in the 1930s: the Depression slowed car sales and unemployment rose.

Figure 13 a and b The Packard Plant in the 1930s (a) and today (b).

Year	Population
1701	City founded
1773	1400
1810	1650
1850	21,000
1900	286,000
1950	1,850,000
2000	951,000
2014	700,000

Figure 14 Detroit population totals.

Phase	Dates	Trends
1	1701–1850	Slow growth
2	1850–1900	Faster growth
3	1900–1950s	Accelerating and very rapid growth
4	1950s–2014	Rapid decline

Figure 15 Detroit population — four phases.

   ISBN: 9780170233316

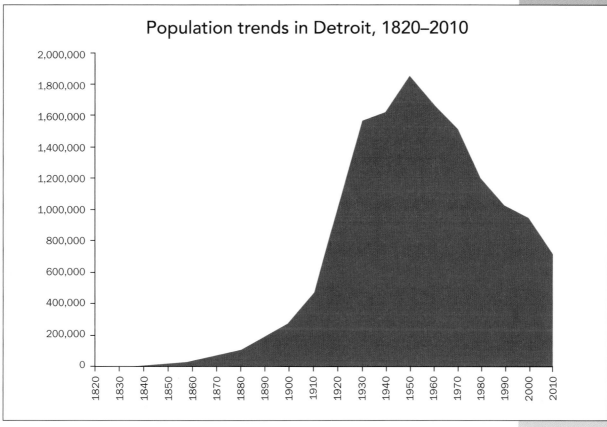

Figure 16 Detroit pattern of population growth and change, 1820–2010.

Comparing other cities with Detroit

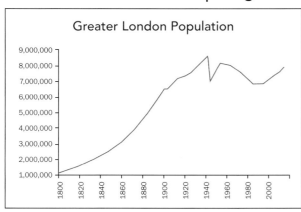

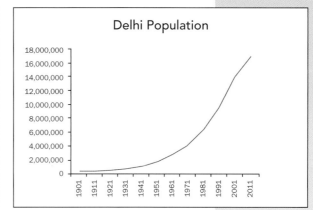

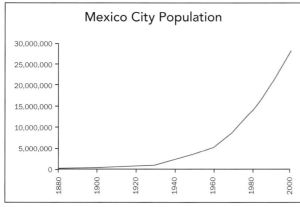

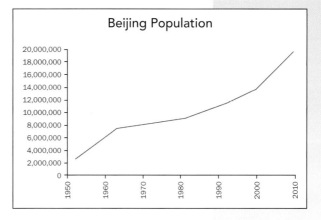

Figure 17 Population changes in London, Delhi, Mexico City and Beijing.

ISBN: 9780170233316

Detroit population facts

Detroit 1850–1950s, rapid growth

- 1850–1950s was a century of growth of business and population in Detroit — entrepreneurs, land speculators and workers migrated to the city from across the world (especially from Europe and Canada) and from other parts of the USA. The city became a magnet for people with skill, ambition and vision.
- People crossing borders transformed Detroit from farmland into an industrial giant. Polish, Lebanese, Irish and Italian immigrants, along with Southern African-Americans (blacks) and whites, arrived in huge numbers during the 19th and 20th centuries to work in auto and other manufacturing plants.
- In 1870, four-fifths of children at school in Detroit had parents born outside of the USA.

Figure 18 Detroit's old City Hall.

![Figure 19 Bustling Detroit streets in the 1950s.]

Figure 19 Bustling Detroit streets in the 1950s.

- Detroit has been described as the Silicon Valley of the early 20th century — a fast-growing and innovative place that offered opportunities for people to get a job and become wealthy.
- In 1914, car manufacturer Ford offered $5 per day for an eight-hour work day. The previous pay rate was $2.34 per day for nine hours. 15,000 workers applied for the 3000 jobs available.
- The success of Ford attracted other auto industry pioneers like Packard, Dodge and Chrysler to Detroit in the early 1900s, just like Hollywood did for movies after 1940 and Silicon Valley did for computers after 1970.
- In 1920, Detroit was the fourth-largest city in the USA. The wealth of the city and its architecture led to it being called the 'Paris of the Midwest'.
- Detroit reached its peak population of nearly two million in 1950. The local population had the highest median income and highest rate of home ownership in the country.
- In the mid-1950s, General Motors, with its headquarters in Detroit, was the largest single employer on earth. Eighty percent of all the worlds' cars were being made in in the Detroit area at this time. Detroit was booming with its thriving auto industry. Many of the residents of the city were auto workers who enjoyed good pay and good benefits.
- But there were signs that all was not well.
  i   Automation in the car plants and competition from foreign car manufacturers were both in their early phases.
  ii Where other major cities in the USA had populations of varied education and skill levels, the auto assembly lines in Detroit required only limited technical ability to perform highly repetitive tasks. The workforce of Detroit was mostly unskilled or semi-skilled. Detroit enjoyed middle-class levels of wealth, but without middle-class levels of skills and education.
  iii African-American (black) workers and their families from the Southern states of the USA had moved north to Detroit in large numbers from the 1930s onwards. They were attracted by the factory jobs and high wages on offer. These African-Americans were welcomed as workers but not as residents by the white population of Detroit. Racism and racial hatred lay close to the surface across the city neighbourhoods.

Detroit population ethnic make-up 1850–1950

Date	White population (%)	African-American (black) population (%)
1850	97	3
1900	99	1
1950	84	16

Figure 20

Detroit 1950s–2014, rapid decline

- The population of Detroit has declined by more than one million people since 1950. The city now is the 18th largest in the USA. The last time the population of Detroit was as low as it is today was in 1910.
- In 1960, average family incomes in Detroit were the highest in the USA. Now, 60 percent of all children in Detroit live in poverty. The unemployment rate is close to 20 percent, twice the average for the USA.
- One-third of all the land in the city is vacant or derelict.
- The average house price in the city is just $NZ30,000. In some parts of the city, houses can be bought for less than $NZ5,000.
- Detroit was hit hard by the 2007–2010 global financial crisis when banks, finance companies and businesses failed. Ford, General Motors and Chrysler cut production as sales fell. All were close to bankruptcy and needed government help to survive. Unemployment rose; house prices fell. Many people in Detroit were unable to pay their mortgages and had homes that were worth much less than they paid for them. Some of these families abandoned their homes and moved away to look for work in other parts of the USA.
- There are 85,000 street lights in the city. The council has switched off most of the lights away from the main roads to save money. Thieves have stripped out the copper wiring from many of the others. Only half the street lights now work.

Figure 21 Decaying Detroit: abandoned homes and empty sections.

- The school system has been rated as the worst in the USA. Close to 50 percent of all the people living in the city have no school qualifications.
- In 2012, there were 411 murders in the city. The murder rate in Detroit is 11 times higher than in New York City. Detroit has been described as the Murder Capital of the USA and the most dangerous city to live in.
- Because of finance problems, the city has cut police numbers by half over the last 10 years. To save more money, most police stations are closed to the public for 16 hours each day.
- More than half of the property owners in Detroit failed to pay their 2011 tax bills. This added to the financial problems facing the city. Public services like schools, libraries and police had their budgets cut.
- A number of council officials have been serving prison sentences for the misuse of city funds.
- The road '8 Mile' that marks the northern boundary of the city has become famous as a marker dividing a black city in decay from more prosperous white surrounding areas.
- Detroit's decline has gained worldwide fame with writers, photographers, artists, reporters and film-makers coming to file reports and images of the city. A French magazine produced a special glossy Detroit issue with models shot in ruined industrial backdrops. It cost US$20, or one-fifth of the price that a house could be bought for in parts of Detroit at the time. Artist Eminem was brought up in the city and has featured Detroit in a lot of his music. He also starred in the movie 8 Mile, set in Detroit.

Detroit population ethnic make-up 1960–2010

Date	White population (%)	African-American (black) population (%)
1960	71	29
1990	22	76
2010	11	83

Figure 22

Learning Activities

1 Geographic concept: Pattern

   a What is the difference between a spatial pattern and a temporal pattern?

   b Give three examples of common types of temporal patterns.

2 a In relation to the Detroit urban area, what is the temporal pattern being investigated in this chapter? (Hint: population total 1700–2012.)

   b What variation to this overall pattern is there? (Hint: population total 1950 to present day.)

3 Study Figures 16 and 17. For each of the five cities, describe how their population has changed over time. In your answer, include some temporal pattern descriptor words like *fast, slow, constant, fluctuating, regular, irregular, accelerating, J curve, peak, trough*; also include specific population totals and dates.

4 a Using the statistics in Figure 14 and the format of the graph in Figure 16 to help you, construct a line graph showing Detroit's population total change in 1701–2014 (full-page landscape orientation would work best).

   b Annotate (attach labels to the graph) in specific places to highlight and explain important features shown on the graph. Use Figure 15 and the text to get information for the annotations.

5 Use the information in 'Detroit population facts' to complete this Q and A activity. You are given the answer and need to write the question for each answer. The first one has been done as an example.

   a 1850–1950s

      i Answer: Migrants.
      Question: What is the name for people who moved from other places like Europe and the South to live and work in Detroit?
      ii Answer: Ford.
      iii Answer: 80 percent.
      iv Answer: African-American.

   b 1950s–2014

      i Answer: 8 Mile.
      ii Answer: One million.
      iii Answer: Unemployment.
      iv Answer: Money.

6 Use the statistics in 'Detroit population facts' to construct three pie graphs showing the ethnic make-up of Detroit's population in 1850, 1950 and 2010. In the 2010 graph, there will be a segment for 'Other ethnic groups', most of whom are Hispanic.

Explaining the population growth of Detroit, 1701–1950s

Important understanding: The population of Detroit grew between 1701 and 1950 due to a combination of natural and cultural factors. Natural advantages and cumulative causation were at work.

There are many reasons to explain why the population of Detroit grew between 1701 and 1950. These factors also explain why growth during the 18th century was slow and why the growth accelerated quickly during the 19th century and during the first half of the 20th century.

Between 1700 and 1800 the area around Detroit was a frontier area and still being explored. Settlement and population growth took off after 1830 as the rich natural resources of the surrounding area were recognised and exploited. Detroit became the city that serviced this area and the people. The big growth trigger during the 1850–1900 period was the Industrial Revolution, with its new technology having an impact across North America. Manufacturing in the Detroit area developed and grew rapidly (Figure 23).

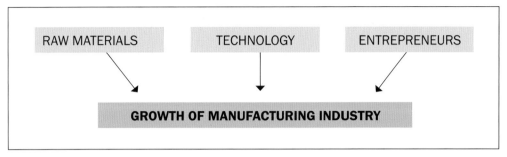

Figure 23 Several factors led to the development of manufacturing.

Factors that led to the growth of Detroit

Detroit had a favourable location partway between the east coast of North America and the interior of the continent (Figure 8). This helped the city to grow as a trading centre that connected the interior of the continent with the east coast and then via the Atlantic Ocean with the rest of the world. Also, the city site next to a large navigable river (Figure 9) gave the city access to the whole Great Lakes waterway system. Canals were built connecting up the Great Lakes with the St Lawrence Seaway, giving Detroit direct access to the Atlantic Ocean and making it an important seaport and trading centre.

The area around Detroit had a lot of resources and raw materials such as timber, copper, iron ore, coal and lead. Water transport as well as overland transport made these raw materials accessible to Detroit. As settlements and farms in surrounding rural areas grew. so the farm produce of livestock and grain began to be shipped through Detroit along with the timber and mineral resources. These natural resources also began to be manufactured in Detroit. Early manufacturing used timber as the raw material to make ships, railway wagons and carriages. After 1850, as the impact of the Industrial Revolution took hold across the USA, business people with money and ideas set up businesses in Detroit. The city became a centre for paint and soap making (chemical industries) and of cooking stove and steam engine manufacture (metal-based industries). Steelworks soon set up in Detroit as well. Detroit for a while in the 19th century was the shipbuilding capital of USA.

The growth of manufacturing in 19th-century Detroit meant the city developed a skilled manufacturing labour force and also became a place of wealth and innovation. A momentum of growth (cumulative causation growth cycle) became established (Figure 25). By the end of the 19th century, Detroit was thriving and the growth momentum continued into the 20th century.

The Ford Motor Company was set up in the city in 1903 by Henry Ford, who was an entrepreneur, i.e. a businessperson with a new idea who was prepared to take on the risk of trying out the idea. The result is financial wealth if the product is successful, but financial ruin if it does not sell. Ford's idea was to mass produce cars on an assembly line. Production of the Model T began in Ford's factory in Detroit in 1913. General Motors and Chrysler soon followed Ford's lead and also began using the assembly line method to manufacture their brand of cars. The city became the auto capital of the world. In the 1930s, Ford alone employed 100,000 workers. Linked industries producing car components like tyres, spark plugs, drive shafts and windscreens set up in and around Detroit (Figure 26).

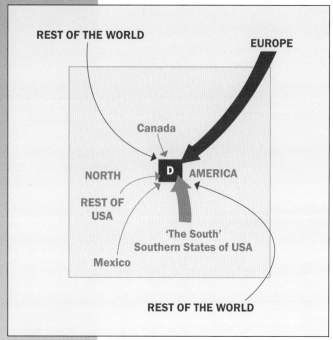

Figure 24 Push and pull factors at work — thousands of people migrated to Detroit between 1910 and 1930.

Workers and whole families migrated to the city in huge numbers from across the USA and the world (Figure 24). Between 1910 and 1920, the population of the city doubled from 465,000 to over one million people. Poverty, unemployment and a struggle to survive in the rural Southern states of the USA and the destruction caused by World War One in Europe acted as push factors for people. Pull factors, especially employment and a chance to become wealthy, attracted thousands of workers to move to Detroit. Detroit was seen as a city of opportunity and a city of prosperity. People saw a move to Detroit as 'a move to the Promised Land'.

The Depression years of the 1930s slowed the growth of the city as factories produced and sold less and unemployment rose. World War Two led to a revival in Detroit as the auto factories switched their production to the making of jeeps, tanks and aircraft for war use.

After World War Two, in 1945 and into the 1950s full production began again in car manufacture and other industries as wealth grew and American families had money to buy the ever-increasing range of consumer products available. The new-model cars rolling off the assembly lines were in big demand. New highways were built across the city and were linked to the fast-expanding national highway network. In the mid 1950s, Detroit was described as 'the most modern city in the world, the city of tomorrow'. It may have seemed this way at the time but the rot soon set in, and the 1950s became a time of transition from a growth city to a decline city.

The process of **cumulative causation** is where growth and development causes more growth and development. One thing causes another thing, which results in growth taking place. This then leads to more growth, and things add up in a cumulative way producing an even bigger result. It is an ever-expanding cycle of growth, with one success leading to more successes.

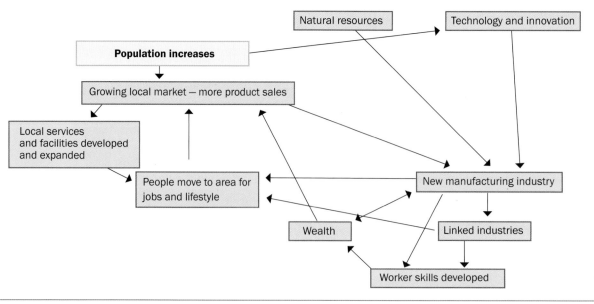

Figure 25 Process of cumulative causation (momentum of growth).

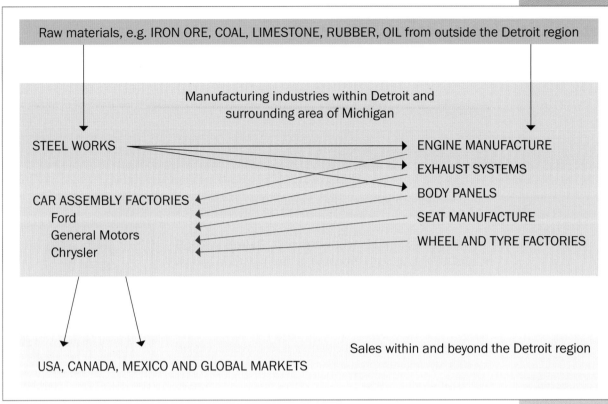

Figure 26 Linked industries in car manufacturing.

Learning Activities

1 **a** Copy Figure 23. For each of the four headings in the diagram, select some key information and examples that are related to Detroit (Entrepreneurs, for example, could include reference to Henry Ford).

b Why would growth in manufacturing cause the population to grow?

c Add a population growth box to the diagram.

2 **a** Make a copy of Figure 24.

b Attach to the diagram a box containing a list of 'Push' factors and another box containing a list of 'Pull' factors.

c Write a short paragraph explaining how 'push-pull' accounts for population growth in Detroit between 1900 and 1950.

3 Study Figure 25 and answer the following questions.

a What does the word 'momentum' mean?

b According to this model, why does population growth and development in an area lead to further and ongoing growth and development?

4 Study Figure 26 and then explain the concept of linked industry. Use examples of the Detroit auto industry in your explanation.

5 **a** Copy the 'Important understanding' from page 136.

b What would you judge to be the most important natural factor and the most important cultural factor that caused rapid population growth in Detroit between 1850 and 1950? Justify your choices.

Explaining the population decline of Detroit, 1950s to present day

Important understanding: People moved away from the city in large numbers as a spiral of decline became established. Some moved to the suburbs outside the city boundary, others moved away from Detroit and Michigan altogether. A combination of push and pull factors worked against the city. A big division appeared, where Detroit city itself was African-American (black), poor and in decline, and the towns and neighbourhoods in areas around the city white, wealthy and growing.

The Motown record company set up in Detroit in 1959 (Figure 27). It quickly became a number one global recording company and record label. Motown music and artists like Marvin Gaye, the Supremes and Stevie Wonder gained worldwide fame for the label, themselves and the city. During the 1960s, Motown music and the city were going in opposite directions. The music was on the rise, while Detroit was a city in transition and going into decline.

Figure 27 Logo of the Motown record company.

Evidence of the decline was easy to see, with the ongoing population loss and urban decay. There were thousands of empty homes, apartment buildings and commercial buildings in the city (Figure 28). The council could not fully maintain city services such as policing, fire protection, schools, rubbish removal, snow removal and street lighting because money was so short. By 2013, the city had run out of money and was bankrupt.

Figure 28 (a – d) Abandoned and decaying factories in Detroit.

Factors that led to the decline of Detroit

The causes of the decline — factories closing, job losses, racial problems and rising rates of crime — have been written about many times. What is argued about is what came first and started the decline.

The big industries began moving away from Detroit to locations with cheaper labour costs and lower taxes from the 1950s onwards. Some of these new locations were in suburbs and towns surrounding Detroit, but many were in other states well away from Detroit (Figure 30). Setting up in a new location also gave the industries the chance to build modern, more efficient factories instead of operating in factories built in Detroit 30 or 40 years ago. In many cases, people followed the jobs. Out-migration and population decline became features of the city. When factories and people moved to areas in the counties around Detroit like Macomb and Oakland, offices, shopping malls, schools, sports arenas and other services were developed to support the middle class and growing populations. These places became self-contained and self-sufficient. Residents had little need to commute to Detroit for work or services.

The auto industry, for a long time the symbol of a modern and prosperous Detroit, has been through a huge slump that has hurt the city. Many car manufacturers have closed down their Detroit factories. In other cases they have kept some production in the Detroit area but have opened new factories elsewhere in the USA, for example in Tennessee and Kentucky, and in Mexico. They have been attracted to these new locations by cheaper labour costs and by the higher outputs that can be gained from high-tech newly built factories. When the car factories close or reduce output, the linked industries suffer (Figure 26). Job losses and unemployment followed by out-migration of workers and their families is often result. The US car producers have been hit by competition from Japanese, Korean and European globally focused auto producers like Toyota, Honda, Hyundai, VW and BMW. These companies produce small and medium-sized, fuel-efficient and green vehicles, which have reputations for good design, safety, economy and reliability. Union power in the Detroit car factories has also helped cause the decline. In the short term the union power in the factories gained higher wages for the workers, but in the longer term this cost workers their jobs as the companies switched to more automated production methods and moved to new locations where they could negotiate lower wages for job security. In the past, African-American migration from the Southern states moved north to Detroit. Now there is a reversal of that trend, with migration from Detroit back to the South taking place.

Three other negatives for the city, which were long present beneath the surface, became noticeable from the 1960s onwards. These were crime, financial mismanagement and corruption by city officials, and racial tension. All three problems are among the 'push factors' that have caused people to leave the city. There are famous markers of these three problems.

Crime

Devil's Night is a name for the night before Halloween (30 October). For a long time this had been a night of minor and mostly harmless mischief and small-scale vandalism across the United States with little major property damage taking place. In Detroit, however, serious vandalism and arson began during the 1970s. These crimes reached a peak in 1984 when there were over 800 fires set across the city. Often it was empty or abandoned buildings that were set alight and destroyed (Figure 29a and b).

Figure 29 a and b Abandoned factories and homes are frequently the target of arsonists.

Financial mismanagement and corruption

Many Detroit city officials have been found guilty of mismanaging and misusing public funds. Kwame Kilpatrick, the mayor of the city between 2002 and 2008, is serving time in prison after being found guilty of obstruction of justice, assault of a police officer, racketeering, tax evasion, extortion and fraud. In 2013, the city declared itself bankrupt.

Racial tensions

The 1967 race riots were the worst in American history and left 43 people dead, 7000 arrested and 3000 buildings destroyed. Reports on the riots pointed to an African-American population that had had enough of discrimination in the workplace where they rarely got promoted to more skilled and better paid jobs, had had enough of living in neighbourhoods with old, rundown and crowded housing where the city spent little money on improving services, had had enough of police brutality, and saw little chance to ever get ahead. The fierce and angry riots were the final straw for many of the better paid (mostly white) residents, and thousands packed their bags and moved to suburbs beyond the city limits and to other states. This 'white flight' was perceived as bad by those concerned about the city finances because of the loss of money (in spending and in taxes) to the city when the wealthier people left and bad by those seeking racial integration. But for many African-Americans, long downtrodden and discriminated against, the 'white flight' was viewed as 'good riddance'.

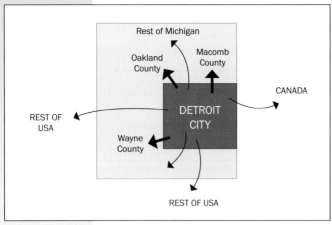

Figure 30 Movement of industry and people away from Detroit, 1960–2014.

The people who remained were those who were too poor to get out. They were also mostly African-American. In the last 40 years the population of the surrounding suburbs has risen from one to over three million, while the city population has shrunk to 700,000. Those who left took most of the jobs, political power and money with them. The city has changed from being 80 percent white and wealthy to be 85 percent African-American and mostly poor. The demography of the suburbs surrounding Detroit, in Oakland and Macomb Counties, is exactly the reverse, with 80 percent white and mostly prosperous middle class (Figure 33). The 2011 census showed a new trend in the outward movement of people from Detroit since 2000 — better off and middle class African-Americans were leaving the city in increasing numbers and moving to more desirable communities in the counties neighbouring Detroit.

ISBN: 9780170233316

Figure 31 Macomb County.

Figure 32 Oakland County.

	DETROIT CITY	WAYNE CO.	OAKLAND CO.	MACOMB CO.
Ethnic make-up				
— African-American	83%	40%	14%	7%
— white	11%	50%	75%	84%
— other groups, e.g. Asian and Hispanic	6%	10%	11%	9%
Average family incomes ($NZ)	39,000	62,000	94,000	79,000
Percentage of families below the poverty line	33%	19%	4%	4%

Figure 33 Detroit City and neighbouring counties — a statistical comparison.

Detroit has suffered a spiral of decline (Figure 34). This is similar to cumulative causation in reverse. Migration away from the city (out-migration) has taken place on a huge scale. The result has been that fewer and fewer people live in the city.

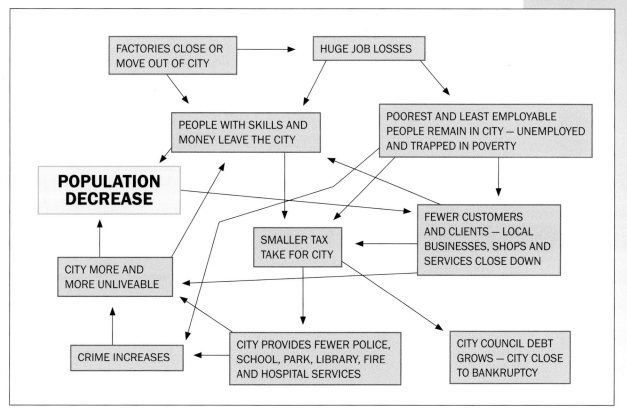

Figure 34 Detroit, a spiral of decline.

ISBN: 9780170233316

Figure 35 Ford global headquarters in Dearborn, Wayne County.

Highland Park, a separate small city completely surrounded by the rest of Detroit city (Figure 35), has become known as the lowest of the low — 'poor, black, burned down and tough'. It was once home to Ford factories with their famous pay deal of $5 a day for workers. In the early 1900s, thousands of families flooded into the city and thousands of Model Ts flowed off the production lines. Between 1910 and 1930, Highland Park was breaking records for huge increases in its population, which rose from 4000 to 50,000. Since 1950, the opposite has happened, and Highland Park has lost 80 percent of its population and now has fewer than 10,000 residents. In 1950, it was known as the 'City of Trees' and a desirable place to live. Ford then moved car production lines and its global headquarters to 'The Rouge' in Dearborn outside of the city boundary. Chrysler remained in Highland Park for longer but it also relocated to the outer suburbs in the early 1990s. Highland Park lost its tax income and had no money. Now Highland Park is the poorest city in Michigan — it has all the problems of Detroit in just 8 sq km. Its entire library system shut in 2002 to save money, and streetlights on residential streets were removed. Residents were asked to leave on their own porch lights to prevent crime. The police force and fire service were left understaffed and using old and out-of-date equipment. According to the fire chief, arson for entertainment and to get an insurance payout took place all the time. Highland Park is a place Detroiters look at as being worse than where they live. A travel guide bleakly warns that *'Highland Park is not highly regarded as a safe city. Visitors should not visit Highland Park at all after dark'*.

Learning Activities

1　**a**　Write down each of these terms and their meaning.

Transition	Decay	Globally focused
Discrimination	Racial integration	White flight
Arson	Corruption	Riot

b　Write a paragraph to explain why the population of Detroit has declined since 1950. Include all the terms above.

2　Copy the diagram and in each of boxes A–E, write headings of factors that have contributed to the population decline of Detroit.

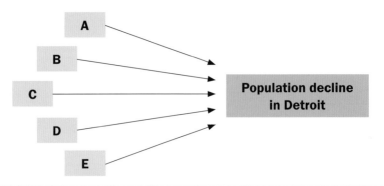

Learning Activities

3 Write one- or two-word captions/labels for each of the photos in Figures 28 and 29.

4 Why did Detroit's once great and flourishing auto industry go into such a decline between 1950 and 2014?

5 Using information about Detroit from the 1950s to the present day, explain how a spiral of decline works and how it resulted in a loss of population for the city.

6 Where is Highland Park and what happened here?

7 Either make a large copy of Figure 34 and then use information from the text to add more information to the diagram to explain and amplify what the diagram is showing (e.g. 'City losing population', 'Movement of people' and 'Prosperous areas in the rest of Michigan'), or turn the statistics in Figure 33 into a set of graphs to compare and contrast demographic features of the four locations.

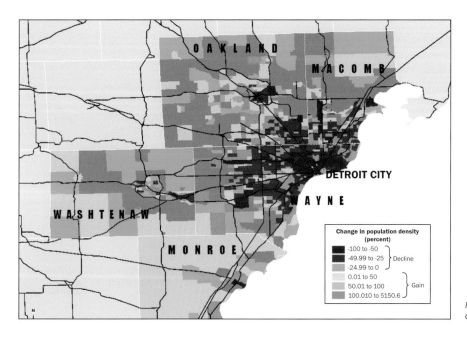

Figure 36 Detroit city and surrounding counties population change 1970–2010.

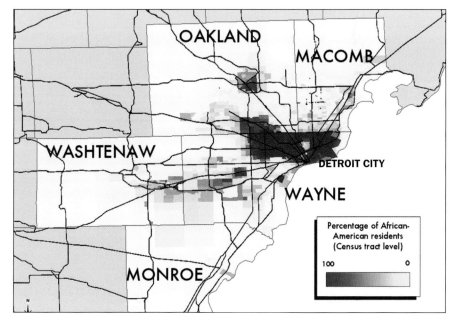

Figure 37 Detroit city and surrounding counties African-American population 2010.

ISBN: 9780170233316

Figure 38 Detroit city area income gap between inner and outer suburbs, 2010.

DETROIT 1950

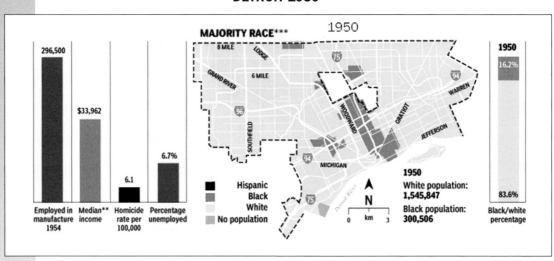

DETROIT 2010

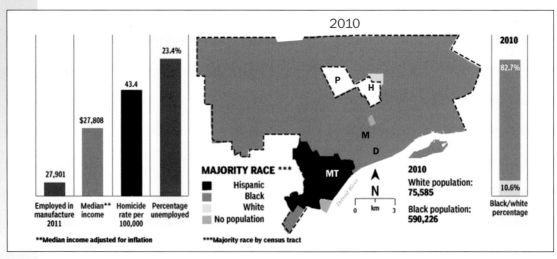

D — Downtown   **Two small cities surrounded by the rest of Detroit city:**
M — Midtown   **P** – Highland Park
MT — Mexicantown **H** – Hamtramck

Figure 39 Infographic showing 60 years of population change in Detroit.

Learning Activities

1 Figures 36, 37 and 38 provide information about the Detroit area (the city and surrounding counties). For each map, describe the spatial pattern shown.

2 Study the infographic in Figure 39. Write a paragraph describing the changes that took place between 1950 and 2010. Here is a start: *'Between 1950 and 2010, major changes took place in the population of Detroit. The number of people who worked in manufacturing (factories) dropped from close to 300,000 to less than 30,000. At the same time, unemployment in the city rose with 23 percent unemployment in 2010, compared with only 6.7 percent in 1950. The ethnic make-up of the city also changed ...'*

The present and the future

Important understanding: Landscapes and places change all the time. Past changes can be recognised and explained, future change can be difficult to predict. People view and perceive Detroit in different ways, some with gloom and despair, others with a sense of excitement and hope.

Figure 40 The inner city of Detroit. Renaissance Centre, Riverfront area.

- The decline and the decay are true but in a large city of 700,000 people there are many positive and hard-working people and good places and pleasant neighbourhoods. Many people who work in factories, shops and offices, in universities, in hospitals and own businesses have well-maintained homes and gardens and well-paid jobs. There are community-minded people trying to make things better by setting up neighbourhood groups to improve the parks, streets and provide activities for the teenagers. These people want the best for themselves, their families and their city.

- The central city Downtown and Midtown areas (Figure 40) have a mix of new and redeveloped high-rise and historic buildings. These are areas of the city with university and hospital neighbourhoods as well as business headquarters. The Renaissance Centre in Downtown is a complex of seven interconnected skyscrapers, one of which is the world headquarters of General Motors. Opening of the centre has resulted in some businesses moving back to the centre of Detroit. Many of these are small businesses set up by IT professionals, artists, designers and musicians. A diversified local economy replacing the dominance of single, large industries may be the future of Detroit. An arts and entertainment hub is beginning to develop in the central area.

- This central area of the city has seen the opening of modern sports stadiums and casinos. These are part of the drive to see Detroit become a major tourism destination. Along the waterfront (beside the Detroit River), a nine-kilometre-long walkway for pedestrians, cyclists and skateborders has been developed (Figure 43). This walkway links some of the new high-rise apartments, office buildings and a cruise ship passenger terminal with parks, marinas and wildlife sanctuaries that are located in the waterfront area. This riverside area has become the home of a variety of annual events and festivals such as auto shows, an international marathon, music and arts festivals.

- Another area of the city that has been revitalised is Mexicantown (Figure 39). This is an area to the southwest of the city centre where large numbers of Hispanic migrants have settled in the past 25 years. Houses could be bought cheaply in neighbourhoods abandoned by many in the white middle class. The area now has thriving businesses and some new housing. Mexicantown buildings have been decorated with bright murals and hand-painted signs, and there are more restaurants per square kilometre than in any other neighbourhood in Detroit. Mexicantown today is one of Detroit's few diverse neighbourhoods, with large numbers of white, black as well as Hispanic residents, and it has been held up as a model of what other neighbourhoods of Detroit could be like in the future.
- In suburban areas with abandoned buildings and vacant land, community gardens and urban farming have been developed.
- 2012-2014 has seen the beginning of a revival for auto companies Ford, General Motors and Chrysler. The companies have restructured. Car sales are increasing and more workers hired. The one remaining car factory in Detroit City, Chrysler, is at full production with its Grand Cherokee Jeep.
- Demographers predict Detroit's population could stabilise in the 650,000 to 700,000 range. The crucial issue, the planners say, is what will happen to the vast suburban neighbourhoods. These are the neighbourhoods people have abandoned, which arc now scarred by crime, blight and poverty. The fear is that these distressed neighbourhoods will continue to wither and decline unless Detroit provides better employment options, better public services, more adult literacy programmes and job training to narrow the gap between the haves and the have-nots.

One vision of the future is of Detroit becoming a smaller, sustainable and model green city with a vibrant and modern Downtown core for businesses and tourists, and with urban farms and pedestrian/bike paths linking this core to compact neighbourhood communities and town centres dotted across the suburban areas (Figure 44).

Figure 41 Wealthy neighbourhoods remain in parts of the city.

Figure 42 Ford Field, in downtown Detroit, is the stadium home of the Detroit Lions American Football team.

Figure 43 Detroit Riverfront development project.

NEIGHBOURHOODS: TODAY AND IN THE FUTURE

The Detroit Future City report suggests that many neighbourhoods in the city now scarred by vacant lots and excessive parking lots could be remade as a 'canvas of green', with trees, parks, farms and ponds filling in the vacant areas.

URBAN MIXED USE

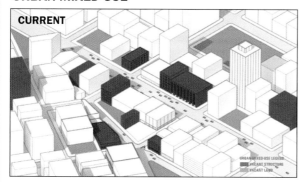

URBAN GREEN

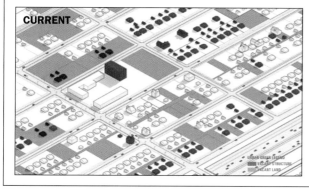

Figure 44 One view of the future Detroit urban landscape.

Learning Activities

Summing up

1 Copy and complete the diagram by outlining briefly in boxes A–D four positive features of present-day Detroit.

2 Name the location where each of the photos in Figures 40–43 were taken and describe what each photo shows.

3 What do planners say is the most challenging issue that Detroit faces?

4 Describe what Detroit 50 years from today could be like, according to Figure 44.

5 *Detroit: an American autopsy* was the title used by Charlie LeDuff for his book about the history of Detroit.

   a What does the word 'autopsy' mean?

   b What makes this an appropriate title for a book about Detroit?

Learning Activities

6 Draw four frames to represent the four cartoon images shown. In each frame, describe the Detroit scene and suggest a date for the image.

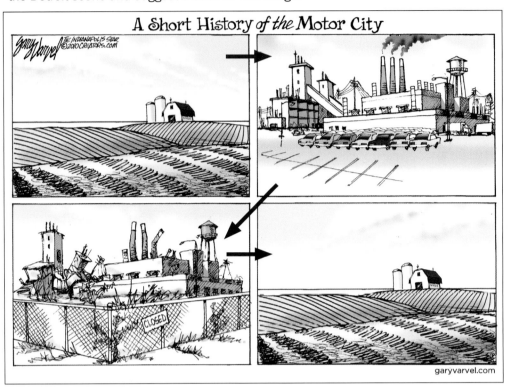

7 Write an essay for each of the following questions.

a Between 1700 and the 1950s, the population of Detroit grew from zero to a peak of two million. A J curve pattern of growth took place — slow between 1700 and 1850, and then accelerating and rapid for the next 100 years. Fully explain this pattern of growth. Hint: give reasons why the city population grew and why this growth was slow at first and then very rapid.

b Since the 1950s, this pattern of growth has been reversed. The population of Detroit has been on a downward curve. Explain fully why this population loss has taken place. Include in your answer the following:

- information specific to Detroit (e.g. names of places, events, people and statistics)

- reference to at least two geographic concepts (some examples are given in the box below)

- visual content, e.g. maps, sketches, diagrams

- this quote from Charlie LeDuff's book: '*Having led us on the way up, Detroit now seems to be leading us on the way down. Once the richest city in America, Detroit is now the nation's poorest*'. (Hint: this could form part an introductory paragraph or part of a conclusion.)

Geographic concepts

- Upward cumulative causation
- Process of migration
- Urban decay
- Location
- Perception

- Linked industries
- Spiral of decline
- Push-pull
- Perspective
- Change

8 Study the infographic shown about the decline (death) of Detroit (page 151).

a Write a list of 10 facts or ideas about the decline of Detroit.

b Design and create your own infographic about Detroit in the growth period between 1900 and 1925.

Learning Activities

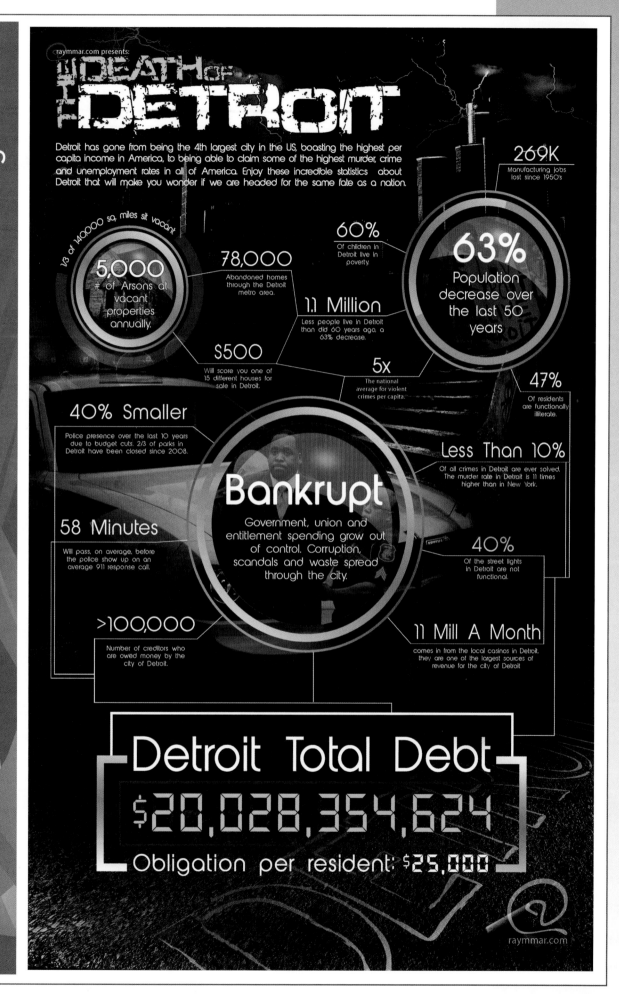

raymmar.com presents:

DEATH OF DETROIT

Detroit has gone from being the 4th largest city in the US, boasting the highest per capita income in America, to being able to claim some of the highest murder, crime and unemployment rates in all of America. Enjoy these incredible statistics about Detroit that will make you wonder if we are headed for the same fate as a nation.

269K
Manufacturing jobs lost since 1950's

1/3 of 140000 sq. miles sit vacant

5,000
# of Arsons at vacant properties annually.

78,000
Abandoned homes through the Detroit metro area.

60%
Of children in Detroit live in poverty.

63%
Population decrease over the last 50 years

1.1 Million
Less people live in Detroit than did 60 years ago, a 63% decrease.

$500
Will score you one of 15 different houses for sale in Detroit.

5x
The national average for violent crimes per capita.

47%
Of residents are functionally illiterate.

40% Smaller
Police presence over the last 10 years due to budget cuts. 2/3 of parks in Detroit have been closed since 2008.

Less Than 10%
Of all crimes in Detroit are ever solved. The murder rate in Detroit is 11 times higher than in New York.

Bankrupt
Government, union and entitlement spending grow out of control. Corruption, scandals and waste spread through the city.

58 Minutes
Will pass, on average, before the police show up on an average 911 response call.

40%
Of the street lights in Detroit are not functional.

>100,000
Number of creditors who are owed money by the city of Detroit.

11 Mill A Month
comes in from the local casinos in Detroit. they are one of the largest sources of revenue for the city of Detroit

Detroit Total Debt
$20,028,354,624
Obligation per resident: $25,000

raymmar.com

ISBN: 9780170233316

Monaco located in Western Europe is one of the smallest countries in the world, yet it is also one of the wealthiest and its citizen enjoy an average income of US$172,000. By contrast, India is one of the world's most populous countries, yet its citizens on average earn only US$1,400 per year.

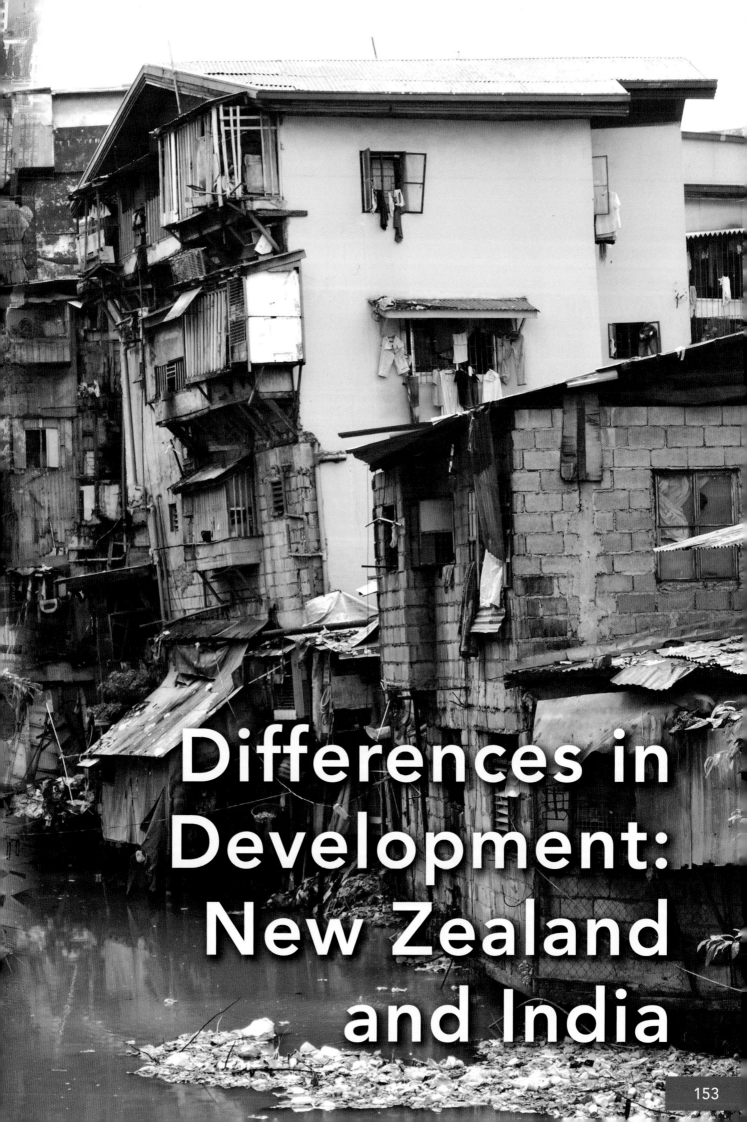

Differences in Development: New Zealand and India

4 Global overview of differences in development

A world of contrasts

Few can deny that we live in a world of incredible contrasts. Of all the world's wealth, 88 percent (or $100 trillion) is concentrated in the few countries that make up Western Europe, North America, Australia and New Zealand. This group of countries representing just 12 percent of the world's total population eat well, consume most of the world's energy, enjoy high incomes and live in excess of 70 years.

Quickly catching up to Western Europe, North America, Australia and New Zealand are the 'Asian Tigers' of South-East Asia. This group, which includes South Korea, Taiwan, Singapore and more recently India and China, all experienced strong economic growth in the latter part of 20th century, bringing wealth, health and prosperity to many in the Asian region.

Figure 1 Luxury homes and yachts in Monaco.

Figure 2 Squatter shacks and houses in a slum urban area.

However, the majority of the world's population is not so fortunate. More than a fifth of the world's population (1.5 billion) have no access to electricity, and a billion more have only an unreliable and intermittent supply. Across most of Africa and in large parts of Latin America, Asia and the Pacific, many people live in poverty. There, large portions of the population lack balanced diets and access to safe drinking water while many are sick and malnourished and go without basic primary education. Others die before they reach adulthood due to armed conflict, disease or a lack of access to basic health care.

 ISBN: 9780170233316

Shown graphically, we can see how the large majority of the world's wealth and income is controlled by just a few. The graph shows that 81.2 percent of the world's total income is earned by just one-fifth of the world's population. By contrast, the poorest two-fifths of the world's population receive just 3.6 percent of all the world's income, which in effect means that they earn (and survive on) less than $2 per day (Figure 3).

Global wealth distribution by population quintiles

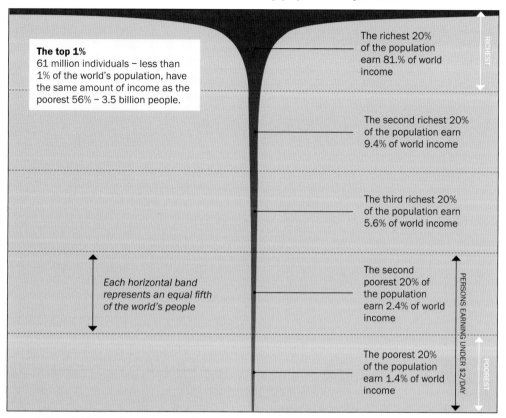

Figure 3 Global distribution of wealth by population.

We begin to get an idea of how the world's income is spatially distributed when it is illustrated on a proportional world map. Of the world's total wealth, 33 percent resides in the United States, while Western Europe has 21 percent, China 10 percent and Japan 7 percent. African countries have the smallest share of the world's total wealth income.

Spatial distribution of income

Figure 4 Distribution of global income.

The characteristics of development

The concept of development

Although it is not difficult to gain an appreciation of the contrasts that exist in our world, the concept of 'development' is harder to define. Traditionally, the concept of development was limited to improvements in wealth and income, however in recent decades it has come to include other aspects of our lives. For example, people now consider access to health and education, democracy and freedom of association as factors that contribute to one's overall quality of life. Therefore, we could conclude that a whole range of factors contribute to the overall 'quality of life' (Figure 5).

Quality of life			
Economic	Physical	Social	Psychological
Income · Job security · Housing, personal mobility	Health · Diet and nutrition, access to drinking water	Employment · Education · Communication networks	Happiness · Freedom · Security

Figure 5 The building blocks of development.

Figure 6 School bus in Korea.

Figure 7 School bus in India.

In Level 2 Geography, the concept of development is defined for us as *the improvement in the quality of life of an identifiable group of people*. The use of the word 'improvement' implies that development could also be regarded as a process (i.e. a sequence of actions that change environments, places and societies).

In this context, development could also be regarded as a strategy to rid the world of poverty. United States President Barack Obama possibly had this in mind when he suggested in 2012 that 'the whole purpose of development is to create the conditions where assistance is no longer needed, where people have the dignity and the pride of being self-sufficient'.

Learning Activities

1   In your own words, define what 'development' means to you.

2   Why is it important that the concept of 'development' encompasses more than just improvements in wealth?

3   Study Figure 5. Of the four main aspects (economic, physical, social and psychological) that make up our 'quality of life', which do you consider the most important? Justify your answer.

Why are some groups of people more developed than others?

Theory 1: Rostow's five stages of development

Many attempts have been made to describe or explain how groups of people move from one stage of development to the next. One of the earliest and widely publicised models to explain economic growth was put forward by economist and political theorist W.W. Rostow in 1960.

Following a study of 15 mainly European countries, Rostow suggested that countries could be placed in one of five stages of economic development: traditional society, pre-conditions for 'take-off', the 'take-off' to self-sustaining growth, the drive to maturity, and the age of mass consumption (Figure 8).

Of the five stages, he proposed that the most important stage in a country's development was the 'take-off' stage, as it marked the transition of a country from a less economically developed country (LEDC) to a more economically developed country (MEDC). Although simplistic, Rostow's main argument was that all countries (regardless of their current level of development) had the potential to break from poverty and develop through five sequential stages.

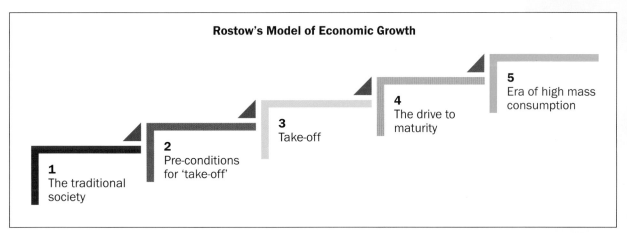

Figure 8 Rostow's five-stage model of economic development.

Learning Activities

1 Using the text as your guide, compile a PMI (Pluses, Minuses, Interesting) summary of Rostow's five-stage model of economic development.

Pluses	Minuses	Interesting points

2 Use the Internet to research the stages of economic development for one of the rich countries listed in Figure 11. Identify the period in the country's economic history that it passed through:

   a the traditional society

   b pre-conditions for take-off

   c take-off

   d drive to maturity

   e era of high mass consumption.

ISBN: 9780170233316

Theory 2: A revised model of development

Rostow's five-stage model does not necessarily reflect the development experiences of all countries, though, particularly poorer ones. Consequently, an alternative model of development has been suggested that divides the countries of the world into four categories, grouped according to their current level of economic development.

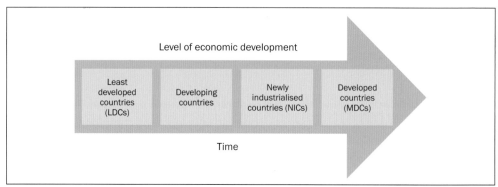

Figure 9 Revised stages of development.

Group 1: In the revised model, least developed countries (or LDCs) are the poorest of the developing countries. Of the 48 countries identified by the United Nations as LDCs, 33 are located in Africa, 14 in Asia-Pacific and one in South America. Together, LDCs account for 12 percent of the world's population but generate a mere 0.9 percent of the world's total output. In other words, one-eighth of the world's population produces less than one-hundredth of the world's total income. Accordingly, LDCs have relatively low incomes per capita, higher rates of poverty, lower life expectancies, and over-dependence on labour-intensive agriculture. The countries of sub-Saharan Africa are regarded the least developed of all the countries of the world. Many of these countries, including Angola, Congo and Sierra Leone, have recently experienced reduced income due to wars, famine and AIDS/HIV despite their significant mineral wealth.

Group 2: Developing countries are, in general, countries that have not achieved a significant degree of industrialisation, and have a medium-to-low standard of living. They are characterised by low incomes and high population growth. Examples of developing countries are Eastern European countries such as Lithuania and Ukraine, and several Middle Eastern countries such as Jordan, Egypt and Iran.

Group 3: Newly industrialised countries, or NICs, which include China, India, Thailand, Brazil, Malaysia, Mexico and others, refers to a group of countries that have in the last 60 years satisfied Rostow's pre-conditions for 'take-off', and have successfully transitioned from a country dependent on farming to one focused on manufacturing.

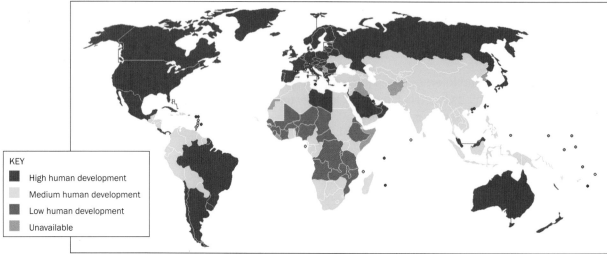

KEY
- High human development
- Medium human development
- Low human development
- Unavailable

Figure 10 Countries with low, high and medium levels of development.

 ISBN: 9780170233316

Group 4: A developed country or 'more developed country' (MDC) has a highly developed economy and advanced technological infrastructure relative to other less-developed nations. They are characterised by high levels of industrialisation and well-developed service and knowledge-based industries (e.g. information generation and sharing, information technology, education, research and development, financial planning). Accordingly, the citizens of developed countries enjoy a high personal income, good health, long life expectancy and a high standard of living.

Thirty-one of the world's most developed countries belong to the Organisation of Economic Development and Cooperation (OECD). Members of the OECD have advanced economies, democratically elected governments, financial stability, strong trade agreements and enjoy extremely high incomes. New Zealand has been a member of the OECD since 1973.

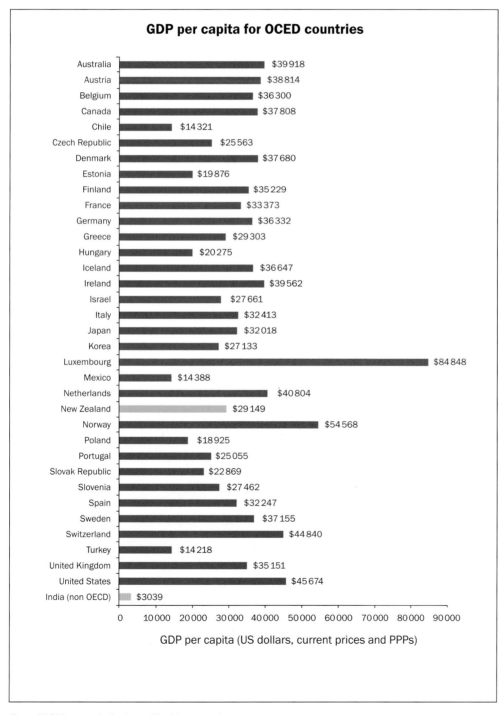

Figure 11 GDP per capita for the world's richest countries.

ISBN: 9780170233316

Theory 3: The North-South divide

The geographical extent of the widening gap between the economic *haves* and *have-nots* was first highlighted in *North-South: a programme for survival* published in 1980. The publication, produced by the Independent Commission on International Development Issues (and nicknamed the 'Brandt Report' after its chairperson, Willy Brandt), identified that differences (or disparities) in global development could be crudely grouped according to the world's two hemispheres — the developed 'North' and the developing 'South' (Figure 12).

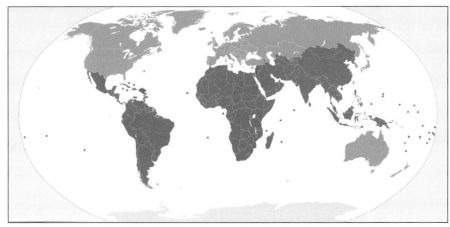

Figure 12 The North-South divide showing the developed 'North' (blue) and the developing 'South' (red).

Theory 4: World Systems Theory and the core-periphery model

Although the North-South division was promoted again as recently as 2001 in the *Brandt Equation*, the economies of the traditional North and South have become increasingly complex. In response, a new way of looking at the pattern of global development has been put forward called World Systems Theory. This new theory suggests that the world's economies (both rich and poor) have been dependent on each other ever since global trade began in the 16th century. As a result, it is now considered more appropriate to consider the world has having three interacting parts:

- the *core*
- the *semi-periphery*
- the *periphery*.

Learning Activities

1 Study Figure 12. Use an atlas or other mapping application (e.g. maps.google.co.nz) to identify at least three countries in each colour category shown on the map.

2 With reference to Figure 12, describe the global pattern of low, medium and high levels of human development.

3 What is the meant by the development 'gap'?

4 Does Brandt's report about the North-South division have any relevance today?

Wallerstein's World Systems Theory model

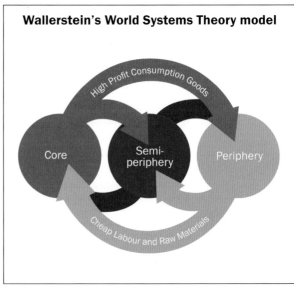

Figure 13 World Systems Theory model.

The *core-periphery model* helps us to understand how the three parts (i.e. core, semi-periphery and periphery) of the World Systems Theory model interact with each other. In the model, the *core* forms the most prosperous and developed part of a country or region. As economic activity and development usually decrease with distance from the core, the *periphery* is relatively poorer and less developed. The growing *semi-periphery* is the part that is currently experiencing the most change as it strives for *core* status. Countries in this group include China, India, Thailand, Brazil, Malaysia, Mexico and Indonesia (Figure 15). Although designed to describe differences in development at the global scale, the core-periphery model works on many scales, from regions to cities, towns to communities.

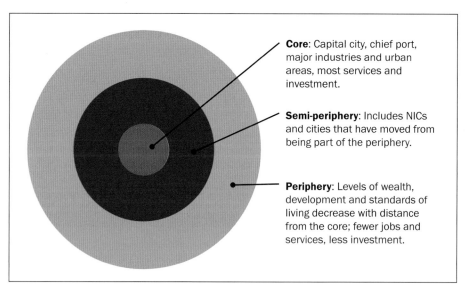

Figure 14 The core-periphery model.

The core-periphery model operates on the principle that as a country or region develops, two processes are likely to occur:

i Economic activity at the core will grow at a faster rate than at the periphery due to its ability to attract new industries and services (e.g. finance, communication and insurance companies). As a result, the core is able to afford facilities such as hospitals, schools, housing and shopping centres, which in turn generate employment opportunities. These facilities and services act as 'pull factors' and encourage inward migration to the core from the periphery.

ii After a period of time, industry and wealth will spread to secondary cores. This can sometimes result in the decline of the original core or result in the development of secondary cores.

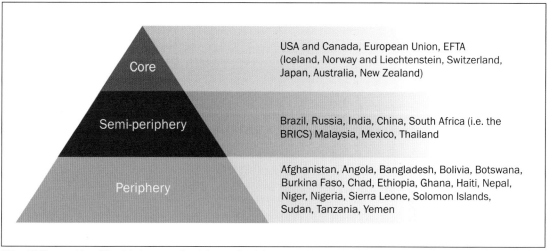

Figure 15 The countries of the global core, semi-periphery and periphery.

ISBN: 9780170233316

Learning Activities

1 Define:

    a core   **b** semi-periphery   **c** periphery.

2 On an outline map of the world, locate and label the 48 countries below currently designated by the United Nations as 'least developed countries' (LDCs).

    a African LDCs: Angola, Benin, Burkina Faso, Burundi, Central African Republic, Chad, Comoros, Democratic Republic of the Congo, Djibouti, Equatorial Guinea, Eritrea, Ethiopia, Gambia, Guinea, Guinea-Bissau, Lesotho, Liberia, Madagascar, Malawi, Mali, Mauritania, Mozambique, Niger, Rwanda, Sao Tome and Principe, Senegal, Sierra Leone, Somalia, Sudan, Togo, Uganda, United Republic of Tanzania, Zambia.

    b Asian LDCs: Afghanistan, Bangladesh, Bhutan, Cambodia, Lao People's Democratic Republic, Myanmar, Nepal, Timor-Leste (formally East Timor), Yemen.

    c American LDC: Haiti.

    d Pacific LDCs: Kiribati, Samoa, Solomon Islands, Tuvalu, Vanuatu.

3 Write a paragraph describing the global distribution of the world's LDCs as listed in Question 2. Your paragraph should identify specific hemispheres, continents and geographic regions.

Development can be measured in a variety of ways

As the concept of development is wide ranging, any attempts to measure it need to extend beyond just measuring economic wealth. A realistic indication of the improvement in the 'quality of life' of a country, region or community should also consider other aspects of life such as physical, social and psychological well-being. In this context, the measurement of people's health, educational attainment and security become very important.

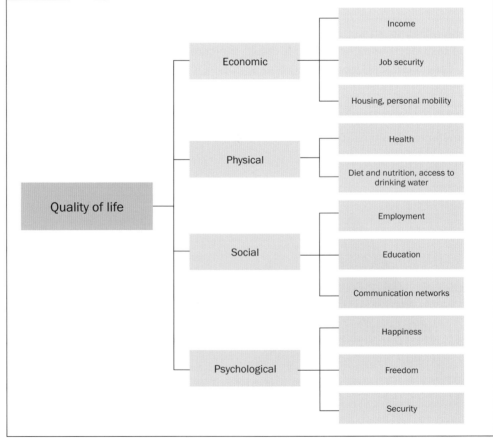

Figure 16 Measurable aspects of development.

Consideration must also be given to the reality that development not only differs between countries but can also vary significantly within countries. Accordingly, most indicators used to measure contrasts between countries can also be used to measure regional differences within countries.

Quantitative versus qualitative indicators

Development can be measured in many different ways, however most measures of development are regarded as either 'quantitative' or 'qualitative':

- Quantitative indicators measure aspects of development that can be counted. In relation to health, examples of quantitative indicators include the ratio of health care professionals relative to the population, or the number of hospital beds per head of population.
- Qualitative indicators are subjective, meaning they are dependent on the perceptions (and biases) of the individual undertaking the measure, and often have a non-numerical basis. Examples include the Corruption Perceptions Index (Figure 31).

Learning Activities

1   How do quantitative indicators differ from qualitative indicators?

2   List the benefits and limitations of using 'quantitative' indicators to measure development.

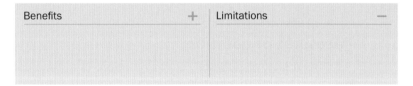

Benefits	+	Limitations	−

3   List the benefits and limitations of using 'qualitative' indicators to measure development.

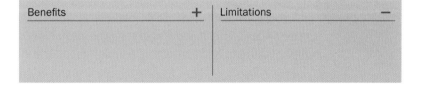

Benefits	+	Limitations	−

Measuring mortality

There are several measures of mortality (death). The most popular measure of mortality is the *crude death rate* (CDR), which is used to quantify the proportional number of deaths in a population in any given year. It is usually expressed as a rate per 1000. Crude death rates are affected by a population's age structure.

Another popular measure of mortality is *life expectancy*. Life expectancy is defined as the average period that a person may expect to live at a given age, however for comparative purposes it is usually calculated at the time of birth. Two points must be kept in mind when using life expectancy as a measure of mortality:

- Life expectancy is an average. Most people live either much longer or much shorter than what the life expectancy indicates.
- Low life expectancy is mostly due to a very high child mortality rate. People who survive the risks associated with childhood can expect to live to a relatively old age, even in countries with a very low life expectancy.

Child mortality, also known as under-5 mortality, refers to the mortality of infants and children under the age of five.

1   Define:

   a   crude death rate

   b   life expectancy

   c   child mortality.

2   What are the limitations of each of the indicators listed in Question 1?

Measuring educational attainment

Quality education is essential to development and is defined as the process of acquiring knowledge, understanding and skills. Two commonly used quantitative indicators of educational achievement are:

- adult *literacy* rate, which measures a population's ability to read and write
- combined primary, secondary and tertiary *enrolment* in schools and colleges.

Improving female literacy and participation in education is currently one of the United Nations development priorities and also one of the Millennium Development Goals (page 200). This is because so many other aspects of development are dependent on improved education. For example, there is a strong correlation between female participation rates in schooling and fertility.

Measures of nutrition

A lack of nutrition, or malnutrition, is a condition whereby individuals are malnourished. It refers to both under-nutrition and over-nutrition. People will be malnourished if their diet does not provide them with adequate calories and protein for maintenance and growth, or when they cannot fully utilise the food they eat due to illness. According to the Food and Agriculture Organization of the United Nations, the average minimum daily energy requirement to be healthy is about 1800 kilocalories (7500 kJ) per person. The World Health Organization (WHO) regards malnutrition as a major health problem, affecting developing countries.

Human Development Index

In recent years, the United Nations (UN) has employed a quantitative measure called the *Human Development Index* (HDI) to gauge each individual country's level of development. The HDI is a composite measure that looks beyond income and considers a nation's 'quality of life' relative to others. The HDI is made up of three dimensions, or variables:

- life expectancy at birth
- an educational index (mean years of schooling, expected years of schooling)
- standard of living (GDP per capita — PPP$).

The three dimensions are each represented by an index value, and these are combined and averaged to give an overall HDI value. The combined index has a maximum value of 1. Every year the UN publishes the Human Development Report (HDR), which uses the HDI to rank all UN member nations according to their 'quality of life' (Figures 17 and 18).

HDI Rank		HDI	Life expectancy at birth	Mean years of schooling	Expected years of schooling	GNP per capita
1	Norway	0.943	81.1	12.6	17.3	47,557
2	Australia	0.929	81.9	12.0	18.0	34,431
3	Netherlands	0.910	80.7	11.6	16.8	36,402
4	United States	0.910	78.5	12.4	16.0	43,017
5	**New Zealand**	**0.908**	**80.7**	**12.5**	**18.0**	**23,737**
6	Canada	0.908	81.0	12.1	16.0	35,166
7	Ireland	0.908	80.6	11.6	18.0	29,322
8	Liechtenstein	0.905	79.6	10.3	14.7	83,717
9	Germany	0.905	80.4	12.2	15.9	34,854
10	Sweden	0.904	81.4	11.7	15.7	35,837

Figure 17 Top 10 countries according to the Human Development Report.

HDI Rank		HDI	Life expectancy at birth	Mean years of schooling	Expected years of schooling	GNP per capita
178	Guinea	0.344	54.1	1.6	8.6	863
179	Central African Republic	0.343	48.4	3.5	6.6	707
180	Sierra Leone	0.336	47.8	2.9	7.2	737
181	Burkina Faso	0.331	55.4	1.3	6.3	1,141
182	Liberia	0.329	56.8	3.9	11.0	265
183	Chad	0.328	49.6	1.5	7.2	1,105
184	Mozambique	0.322	50.2	1.2	9.2	898
185	Burundi	0.316	50.4	2.7	10.5	368
186	Niger	0.295	54.7	1.4	4.9	641
187	DR Congo	0.286	48.4	3.5	8.2	280

Figure 18 Bottom 10 countries according to the Human Development Report.

Learning Activities

1. What advantages do composite indices (e.g. HDI) have over single-measure indicators (e.g. GDP per capita)?

2. Name the three variables used by the United Nations (UN) to calculate the overall HDI value for a country. Justify the use of each of the three variables.

3. Imagine that the UN has decided to add one more measure of development to the HDI. Discuss with your class what you think it should be and why.

4. What do the countries in Figure 18 have in common in terms of their global location?

5. Construct a pie graph to illustrate the following table of data:

Level of human development	HDI value	Number of countries
Very high	0.9 and over	38
High	0.80–0.89	45
Medium	0.50–0.79	75
Low	Below 0.50	24

6. Visit the UN website hdr.undp.org and go to the statistics section to see how the HDI has changed over time. The website gives HDI data for each country, organised into high, medium and low HDI. Select one country from each of the three categories and compare them using several indicators. Present your findings to your class.

Quality of Life Index

The *Quality of Life Index* differs from other indicators in that it employs a combination of both quantitative and qualitative measures to determine an overall 'quality of life'. The composite indicator uses nine 'quality of life' factors to determine a nation's score. For 2013, the top 15 countries and their scores according to the quality of life survey were:

1	Switzerland	215.71
2	Germany	204.84
3	United States	199.56
4	Sweden	191.36
5	Canada	186.03
6	United Arab Emirates	186.01
7	Denmark	182.29
8	Norway	173.86
9	Qatar	169.92
10	Austria	167.39
11	Finland	167.21
12	Australia	165.80
13	New Zealand	163.17
14	Netherlands	160.54
15	Japan	159.79

Learning Activity

1 Go to Gapminder World (www.gapminder.org) and study the wide range of interactive maps about the wealth and health of nations. Compare two or more quantitative measures of development. Share your findings with the class.

Factors contributing to differences in development

There have been numerous discussions about which factors most contribute to differences in development. What is clear is that variations between countries are most likely the result of a combination of both natural and human factors, as shown in Figure 19.

Natural factors	• Landlocked and small island countries are at a considerable disadvantage and have generally developed more slowly. • Countries with a tropical climate have developed more slowly than those with a temperate climate due to less productive agriculture. • Some countries have a wealth of natural resources which have been harnessed to spur economic activity.
Human factor: economic	• Free market economies that are open to trade an foreign investment have developed faster than closed economies. • Faster growing countries tend to have higher rates of saving and investment.
Human factor: social and political	• Countries with democracy, stable government, law and order, and minimal corruption have more advanced levels of development.
Human factor: demographic	• Youthful populations limit a country's ability to develop. • The highest rates of growth are experienced in countries with low birth rates and longevity.

Figure 19 Factors contributing to differences in development.

 ISBN: 9780170233316

Another approach proposes that the factors that contribute towards a country's level of development will usually fall into one of three spheres: factors contributing towards social-cultural well-being, factors contributing towards economic well-being, and factors contributing towards environmental well-being **(Figure 16)**. Each of the three factors can be further subdivided into several specific areas:

- Factors that contribute towards a social-cultural well-being of the population, for instance, include the level of participation and attainment in education, access to training and educational facilities, as well as physical and mental health, and one's overall sense of well-being.

- Population growth and economic development put pressure on the sustainability of the environment and affect a country's overall economic well-being.

- The quality of the natural and built environment is directly related to people's environmental well-being. Physical aspects of the natural environment that have a substantial impact on the quality of life include air, soil and water quality, access to safe drinking water and waste disposal.

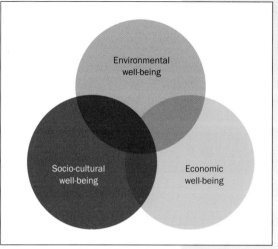

Figure 20 The three spheres of development.

Strategies for reducing development inequalities

The World Bank, an international institution that provides loans to developing countries, identifies five strategies that successfully developed countries have all implemented in common:

1 **Economic growth:** Countries that have reduced poverty substantially and in a sustained manner are those that have grown the fastest. Successful development requires sustained periods of high per capita income growth.

2 **Promote a vibrant private sector:** Private firms, including small and medium-sized businesses in rural non-farm areas, play a critical role in generating employment, particularly for youth and poor people.

3 **Empowerment:** All people should have the ability to invest in their health and education, to shape their own lives by being able to participate in the opportunities provided by economic growth, and to have their voices heard about decisions that affect their lives. Access to essential public services such as health, education and safe water is critical and should be provided equitably.

4 **Stable governance and an efficient market:** A democratic and stable government fosters a society where the public sector facilitates the private sector, contracts are enforced and markets can operate efficiently. It also ensures that basic infrastructure (e.g. energy supplies, transportation and telecommunication) function, health and education services are provided, and people can participate in decisions that affect their lives.

5 **Ownership:** Countries need to own their development agenda. This helps ensure that there is widespread support for development programmes and the reform measures that underpin them.

Theme 1: New Zealand's place in the more developed world

New Zealand in profile

Why is New Zealand considered a developed country? To answer this question, one only needs to compare our 'quality of life' with that of the rest of the world. For example:

- New Zealand's small population of 4.4 million enjoys a 'quality of life' much higher than that of 95 percent of the world's population.
- New Zealand performs exceptionally well in most measures of well-being and ranks among the top countries of the OECD in a number of areas.
- New Zealanders on average earn US$18,601 a year. Although this is less than the OECD average of US$22,387 a year, it is considerably higher than the global average. Accordingly, the people of New Zealand have the means to enjoy a high standard of living.

Land area (sq km)		270,534
Population		4,447,922
Natural increase rate (%)		1.2
Infant mortality (per 1000 live births)		4.7
Life expectancy (years)	Male:	78.8
	Female:	82.7
Doctors per 1000 population		2.3
Literacy rate (%)	Male:	99
	Female:	99
GDP per capita (US$)		29,813
Gini coefficient		36.2
Population living in poverty (%)		0.1
Key industries	Food processing, wood and paper products, textiles, machinery, transportation equipment, banking and insurance, tourism, mining	
Labour force by sector (%)	Agriculture:	7
	Manufacturing:	19
	Services:	74
Land use (%)	Arable land:	5.5
	Permanent crops:	6.2
	Other:	87.6

Figure 21 New Zealand's statistical profile.

 ISBN: 9780170233316

It is no surprise, then, that New Zealand's Human Development Index (HDI) measure is very high relative to the rest of the world. In fact, New Zealand's HDI measure of 0.908 puts it fifth place in the world and only behind Norway (first), Australia (second), Netherlands (third), and USA (fourth) (Figure 17). To understand why New Zealand compares well on a global scale, one needs to examine how well New Zealand does in each of the component measures that make up the HDI: health, education and income (Figure 22).

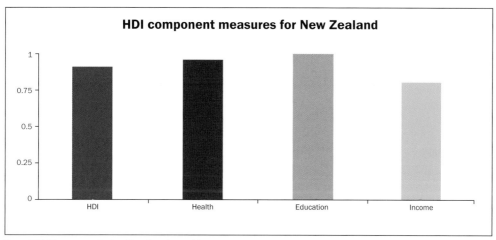

Figure 22 HDI components: health, education and income.

Despite New Zealand's strong showing in HDI measures, its wealth is not evenly distributed throughout its population. More than half of the country's wealth is owned by just 10 percent of the population, while the poorer half of New Zealand's population controls less than 5 percent of the nation's total wealth.

Overall, the 'quality of life' in New Zealand continues to improve over time, with increases in life expectancy, household income and improvements in health and safety.

Factors contributing to New Zealand's level of development

Factor 1: Historical development

An analysis of New Zealand's development through time reveals that our nation has been able to transition from that of a traditional society described in Rostow's model (page 157) to that of a society of high mass consumption in a relatively short period of time.

New Zealand's development began with the arrival of early Maori. Prior to European colonisation, the economic undertakings of Maori were largely limited to the utilisation of resources in important resource areas. Economic activities mainly involved the gathering of kai moana (seafood) and the hunting of various bird species, most notably kereru, kaka, kiwi and weka. Crops were cultivated as Maori became more established, starting with exotic kumara (sweet potato), taro and hue (bottle gourd) introduced by the first settlers and followed by the cultivation of native karaka berries and ti kouka (cabbage tree). Pre-European trade took the form of gift exchange, with inland tribes often trading foods with coastal tribes. Stone (notably pounamu) was traded throughout the country.

A little over 170 years since colonisation, New Zealand now finds itself at the forefront of global economic development. However, New Zealand's global status and relative wealth has not come about in that time through chance. A number of factors have in different ways contributed to New Zealand's level of development and the way of life New Zealanders enjoy (Figure 23). Unfortunately, not all New Zealanders have benefited from New Zealand's economic growth. This is because the fruits of New Zealand's economic development have not always been distributed uniformly or evenly throughout the country.

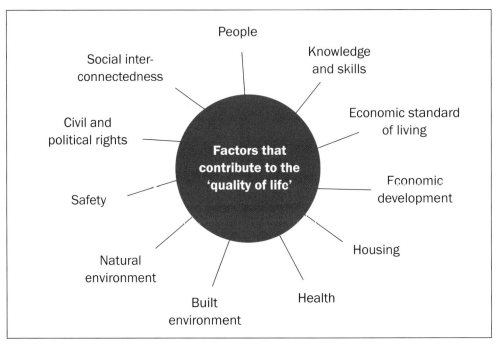

Figure 23 Factors that have contributed to New Zealand's 'quality of life'.

Factor 2: Population structure

The shape of a nation's population pyramid tells us a lot about its past, its present and its projected development. Its shape reflects the influence of births, deaths and migration on its population over time, and shows whether its population is expanding, stable or likely to decline. In general, a population with a high birth rate and low death rate will have a broad-based, triangular-shaped pyramid, while populations with low birth rates and low death rates are usually narrower at the base and have straighter sides. The shape of New Zealand's population pyramid reflects that of the latter in that a smaller proportion of its population is in the pre-reproductive age groups and a larger percentage is in or about to enter the post-reproductive age groups (Figure 24). This suggests that New Zealand has an ageing population.

In 2013, 20 percent, or one-fifth, of New Zealand's population was under 15 years of age, and 13 percent were over the age of 65. This indicates that New Zealand has low birth, infant mortality and death rates, and long life expectancy. This, in time, will cause the numbers entering the reproductive age groups to decline, leading to a decline in the total population.

However, birth and death rates and life expectancy are not constant among all people groups within New Zealand. For instance, the European population is the oldest, with a median age of 38 years, followed by the Asian population at 28 years, the Maori population at 23 years, and Pacific peoples at 22 years. Knowing the differences and similarities between groups within a population helps us to understand the diversity of needs that make up a society. It also assists decision-makers to anticipate potential pressures on the wider social, economic and physical environment.

Factors such as population location, growth, age, ethnicity and population movement all contribute to a range of issues affecting the quality of life. For instance, more than half of New Zealand's population (56 percent) live in its 12 largest cities, which are projected to account for 86 percent of New Zealand's total population growth by 2026, with two-thirds of the growth occurring in the Auckland region. About half of this growth is expected to be the result of internal migration, and the other half will be made up of new immigrants to New Zealand who are attracted to Auckland due to its 'gateway' city status. Accordingly, Auckland has a higher ethnic diversity than other regions of New Zealand and has higher proportions of Asian and Pacific Island peoples compared with the rest of New Zealand.

New Zealand cities tend to have a lower median age than the national median (35.9 years). This is in part due to the higher proportion of people in tertiary (e.g. university or polytechnic) training or of working age. South Auckland and Porirua have the youngest populations, with more than a quarter of their population under 15 years of age.

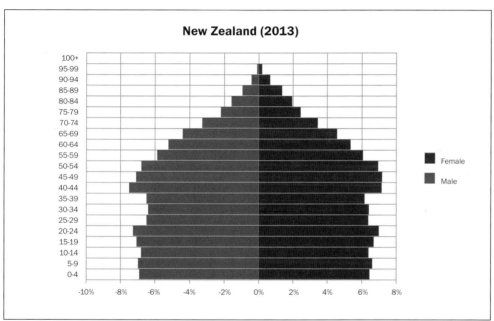

Figure 24 New Zealand's age-sex structure (2013).

Factor 3: Health

Age, ethnicity, socio-economic status, employment, education and housing as well as external factors such as living conditions and access to health services and recreational opportunities all impact on the overall physical and mental health of populations.

Various measures and indicators show that the overall health of New Zealanders has improved over time. However, significant differences in health and well-being do exist between different population groups within New Zealand:

- The life expectancy of New Zealanders overall is increasing, although male life expectancy still lags that of females. Life expectancy for Maori is lower than that for non-Maori. In socio-economic terms, there is a nine-year difference in life expectancy for males at birth between the richest and the poorest tenth of the population.
- Nationally, there has been a decline in the rate of general practitioners (doctors) per 100,000 population. Auckland has the highest rate of general practitioners; the Rodney area north of Auckland has the lowest.

Factor 4: Housing

The need for shelter is a fundamental component of the 'quality of life'. Without it, people cannot meet their basic needs and cannot adequately contribute to society. Poor-quality housing can affect people's health, education and community well-being.

Figure 25 High-quality housing.

Figure 26 Low-quality housing.

Changing demand for housing can put pressure on both the natural and social environment and can affect society's ability to provide suitable infrastructure and services. In New Zealand, the proportion of people who own their own home is decreasing. This decline can be attributed to a number of factors, including increasing property values and the cost of home ownership, and the increasing affordability of renting relative to household incomes.

Factor 5: Economic standard of living

Levels of income and wealth directly influence individual or family well-being, and collectively the prosperity of the entire population. The more affluent an economy, the better off its people are in terms of opportunities to gain a better income, material goods and services such as specialist health care and higher education, which in turn can lead to greater social cohesion, educational advancement, diverse employment opportunities and increased life expectancy.

While incomes have increased markedly in New Zealand over the past decade, differences in income still exist. For instance, the top 10 percent of wealthy individuals own more than half of the nation's total net worth, while the bottom 50 percent of the population shares just 5 percent of total net worth.

Income differences also exist in New Zealand on a regional scale. In 2012, the highest average weekly incomes were found in Wellington ($791) and Auckland ($765), and the lowest in Northland ($590), Manawatu-Wanganui ($632) and Gisborne-Hawke's Bay ($652) regions.

Factor 6: Natural environment

New Zealand is a mineral-rich country, with a large variety of mineral deposits including natural gas, gold, iron ore, sand and coal, and has an abundance of timber, hydropower and limestone. Its resource wealth has been extracted, harnessed and developed over the last 170 years to directly and indirectly provide income to New Zealanders, in turn raising the nation's the standard of living far above that of most of the world.

However, in recent decades, demand for New Zealand's resources has been driven by economic growth in developing countries such as India and China. Left unchecked, this demand has the potential to threaten the quality of our unique natural environment and directly harm our quality of life. Economic development and population growth, which can lead to the expansion of urban areas (urban sprawl), places pressure on the sustainability of the natural environment.

Furthermore, problems such as environmental pollution, waste generation and management, heritage protection and the protection of natural ecosystems on both land and sea must be managed in a sustainable way to ensure that our use of the environment guarantees the well-being of future generations. Recognising the need to balance economic development with its environmental responsibilities, the Government's goal is for New Zealand to 'make the most of its abundant energy potential through the environmentally responsible development and efficient use of the country's diverse energy resources'.

Figure 27 New Zealand is rich in minerals.

Figure 28 Economic development places pressure on the sustainability of the natural environment.

Factor 7: Safety

The right to feel safe and secure in our homes and community is a basic human right and a pre-condition to the overall health of the community. However, as Figure 29 shows, perceptions of safety vary considerably throughout New Zealand.

Urban area	A bit unsafe	Very unsafe	Totally unsafe
Rodney	19.1	8.5	27.6
North Shore	21.9	7.7	29.6
Waitakere	30.7	16.6	47.3
Auckland	27.5	12.2	39.7
Manukau	25.6	22.5	48.1
Hamilton	33.7	13.4	47.1
Tauranga	29.9	12.7	42.6
Porirua	21.5	12.1	33.6
Lower Hutt	24.9	12.1	37.0
Wellington	17.0	4.5	21.5
Christchurch	27.3	13.2	40.5
Dunedin	22.1	5.2	27.3
New Zealand	23.6	11.2	34.8

Figure 29 Percentage of people aged 15 year and over who reported they felt unsafe walking alone in their neighbourhood at night.

Factor 8: Civil and political rights

Human development is defined not only in terms of freedom from hunger and poverty but also in respect to the dignity of the individual. For many countries of the world, this ideal is preserved through the freedoms associated with the concept of democracy.

There are two distinct aspects of democracy: electoral competition, which determines who will be in power; and the various checks and balances of a democratic society that determine how that power is used, such as the freedom of speech and of the media, and the right to free association (e.g. the right to belong to interest groups, political

ISBN: 9780170233316   PHOTOCOPYING OF THIS PAGE IS RESTRICTED UNDER LAW.

parties, lobby groups and trade unions). Combined, the two aspects of democracy help to ensure that elected decision-makers are honest and subject to electoral accountability.

In this context, democracy matters because it reflects an idea of equality, freedom and that each individual should have an equal vote and an equal say in the formation of their government. New Zealand is a democracy, meaning New Zealanders have ultimate power over the way they are governed. The democratic rights of New Zealanders include:

Electoral rights, the right to take part (directly or indirectly) in government, and the right to equal access to the public service. There is an associated duty of responsible citizenship, being a willingness to play one's part in public affairs and to respect the rights and freedoms of others. Taken together, these rights ensure the ability to participate in public and political life.

(Universal Declaration of Human Rights)

Should begging be banned?

A crackdown on beggars in Auckland has the support of police, who believe it could reduce crime and anti-social behaviour. In a submission to the Auckland Council on the review of bylaws on public safety, Inspector Gary Davey, Auckland City police prevention manager, said police supported stricter changes in some areas. The begging issue is being debated by council, with an inital draft of the bylaw banning asking for money or food

Figure 30 Homeless street beggar sits on Queen Street, Auckland.

'in a manner that may intimidate or cause a nuisance'. After public feedback, commissioners appointed by the council then recommended all begging be banned.

Learning Activities

1 Read the news article 'Should begging be banned?' Do you agree that begging should be banned?

2 Debate as a class whether begging should be considered a social issue or a crime.

Freedom from the corruption of public officials is at the heart of a relatively modern measure of honesty called the Corruption Perceptions Index. The index ranks countries and territories based on how corrupt their public sector is perceived to be (Figure 31). The latest survey suggested that, along with Denmark and Finland, New Zealand is perceived to have the least corrupt public sector in the world. By contrast, India is ranked 94th.

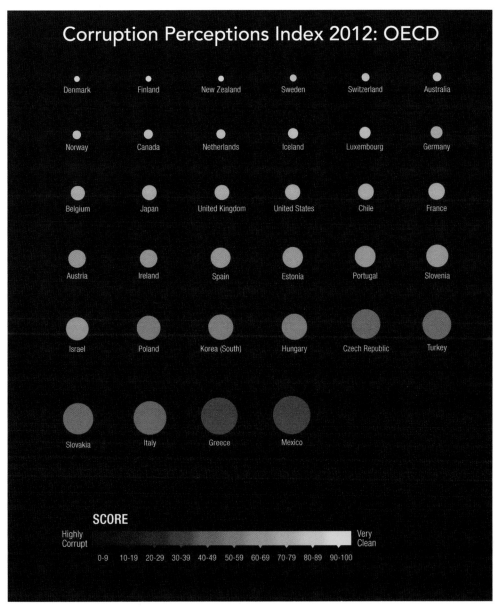

Figure 31 Corruption Perceptions Index.

Factor 9: Education and skills

Knowledge and skills improve people's ability to meet their basic needs, widens the range of employment options available to them and enables them to influence the direction their lives take. Furthermore, research shows that the skills people acquire through schooling and ongoing on-the-job training also enhances their sense of self-worth, security and belonging.

New Zealand's resident population is highly skilled, although regional differences do exist.

- **Early childhood education:** Research shows that 'the early childhood years are vital to a child's development and to their future ability to learn'. New Zealand performs very well in this area, with over 92 percent of three year olds and nearly 100 percent of four year olds enrolled in early childcare education services such as kindergartens, play centres and te kohanga reo.
- **School leavers:** More than 70 percent of school leavers in New Zealand leave with the qualification of NCEA Level 2 or above. Evidence suggests 'those who leave school early with few qualifications are at a much greater risk of unemployment or vulnerability in the labour force of having low incomes'.

Figure 32 Clocktower of University of Otago, Dunedin.

Figure 33 Hagley Park, Christchurch.

Figure 34 Starship Children's Hospital.

• **Tertiary education** Recent years have seen an increase in tertiary education enrolment from 8.4 percent in 1998 to a peak of 14 percent in 2005. This is relevant, as acquiring a tertiary qualification provides individuals with additional skills and knowledge to participate in society and in the economy. Interestingly, females (14 percent) are more likely than males (11 percent) to participate in tertiary study.

An individual's level of educational attainment is closely related to their income level. At the last census, annual personal incomes were highest for those whose highest qualification was a doctorate degree ($69,900 per year) and lowest for those with no qualifications ($16,900 per year).

Factor 10: Built environment

The built environment refers to the human-made surroundings that provide support for human activity. It includes buildings, parks, neighbourhoods and cities and its supporting infrastructure such as water supply, telecommunications and energy networks.

The quality of the built environment affects people's access to facilities and services and contributes to the way people feel about where they live. For instance, poor access to facilities and services can impact on a person's health, their financial well-being and their sense of safety. It also impacts on the sustainability of the natural environment.

In general, most New Zealanders are proud of their community's built environment, and in urban areas regard it as easy to access:

• high-quality health services
• a range of primary and secondary schools, and high-ranking tertiary institutions
• public transport facilities and consider it affordable, safe and convenient
• a local park or green space.

Factor 11: Social connectedness

Social connectedness refers to the relationships people have with others and the benefits these relationships can bring to the individual as well as to society. People who feel socially connected are inherently happier than those who do not have the relational support of family, friends, colleagues and neighbours. Accordingly, feelings of isolation and loneliness can affect overall well-being and can be harmful to people's physical and emotional health, resulting in stress, anxiety or depression. Perceptions of happiness vary only slightly across New Zealand, as illustrated in Figure 35.

 ISBN: 9780170233316

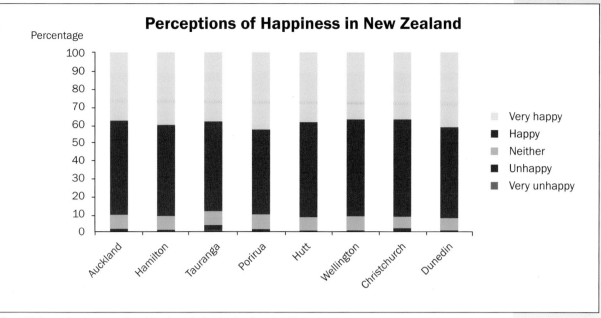

Figure 35 Comparing the perception of happiness in New Zealand's major cities.

1 Why is New Zealand considered a 'more developed' country? Support your answer with quantitative evidence from the text.

2 What challenges does New Zealand's ageing population present with regard to its future development?

3 Construct a percentage bar graph for each of the ethnicities represented in the income table. Comment on the patterns illustrated in each graph.

Income	European (%)	Maori (%)	Pacific peoples (%)	Asian (%)
Less than $20,000	39.54	42.78	36.49	52.82
$20,000–$50,000	36.67	36.68	40.73	29.1
Over $50,000	18.55	9.05	5.87	9.16
Not stated	5.24	11.49	16.92	8.92

4 Construct a multi-line graph using the data presented in the table. Compare and comment on the trends illustrated in the graph.

Year	No qualification	School qualification	Post-school and school qualification	Post-school but no school qualification
1986	566,600	373,800	458,400	207,800
1988	496,300	376,800	503,900	178,600
1990	407,400	398,200	557,100	168,600
1992	383,100	397,300	594,000	149,300
1994	401,500	429,000	663,100	161,300
1996	381,500	470,600	744,700	165,700
1998	368,700	459,900	766,000	163,100
2000	357,500	462,500	859,600	164,100
2002	366,600	490,400	887,800	182,600
2004	405,000	506,800	962,600	195,200
2006	407,900	469,300	989,600	226,600
2008	371,600	507,200	1,051,900	202,400

Learning Activities

5 Using the data in the table and an outline regional map of New Zealand, construct a proportional symbol map to illustrate differences in weekly income across New Zealand. Write a brief paragraph describing the pattern shown in the map.

Region	Average weekly income ($)
Northland	543
Auckland	727
Waikato	640
Bay of Plenty	608
Gisborne/Hawke's Bay	615
Taranaki	702
Manawatu-Wanganui	578
Wellington	789
Tasman/Marlborough/West Coast	671
Canterbury	598
Otago	672
Southland	710

6 Rank the following factors that have contributed towards New Zealand's level of development from most influential to least influential. Write a paragraph justifying your ranking.

Population structure and composition	Health	Quality of housing	Personal income
Natural environment	Safety and security	Civil and political rights	Education and skills
Built environment	Social connectedness	Economic standard of living	Heritage

7 New Zealand has transitioned from that of a traditional society to one that is characterised by high mass consumption in a relatively short period of time. Using evidence from the text, outline the differences in development that have resulted in the areas of:

a population structure (e.g. life expectancy)

b health

c housing

d income (i.e. the economic standard of living).

8 Go to the Statistics New Zealand website (www.stats.govt.nz) and access the site's Interactive Boundary Maps tool, which shows facts about the geographic areas in New Zealand. Use the web-based tools to search the range of 'Quickstats' to explore disparities between the different geographic regions of New Zealand.

Strategies for reducing development inequalities

Strategies for reducing the development inequalities refer to the actions that may be taken by governments and non-governmental organisations to reduce disparities. This suggests that inequality is not inevitable and that governments and non-governmental organisations can do something about it. The OECD, of which New Zealand is a member, recommends three strategies for reducing development inequalities:

1 Reform tax and welfare policies
2 Lower unemployment
3 Increase accessibility to high-quality public services.

Strategy 1: Reform tax and welfare policies

Despite growth in the average take-home income, inequality among working-age people in New Zealand has increased significantly over the past 25 years. This is because:

- Over that time, household incomes have grown by 2.5 percent per year for the richest 10 percent of New Zealanders, but only by 1.1 percent for the poorest 10 percent.
- The share of the nation's total income of the top 1 percent of income earners has risen from 6 percent in 1980 to 9 percent in 2005, and that of the top 0.1 percent of earners has more than doubled from 1.2 percent to 2.7 percent. At the same time, top marginal tax rates declined from 60 percent in 1980 to 33 percent in 2010.

Adjusting tax and benefit policies is the most direct strategy available to Government for redistributing income. New Zealand's experience demonstrates that income inequality is largely market driven. Between the mid-1980s and mid-1990s, market-driven income inequality increased sharply in New Zealand, and the tax and benefit system were seldom used to counter the increase. As a result, the tax share of household income declined greatly, increasing household disposable income, especially for the higher-income households.

However, since the introduction the Working for Families system of tax credits in the mid-2000s, income inequality has slightly decreased in New Zealand. Introduced in 2004, the policy, which provides extra money to almost all families with children earning under $70,000 a year, saw income inequality and poverty fall in New Zealand for the first time in two decades.

Strategy 2: Lower unemployment

Implementing strategies that help a country move towards full employment is another effective way of reducing inequality. However, the challenge of creating more and better jobs that offer good career prospects and a real chance for people to escape poverty is not necessarily a simple one for governments. Upskilling the school leaver and working population is key to lowering unemployment. This must begin from early childhood and continue through secondary and tertiary education. Once the transition from school to work is complete, there must be sufficient incentives for workers and employers to upskill throughout their working life.

It is also important that the academic areas that tertiary graduates are graduating in are aligned with the needs of the economy, both spatially and in response to skills shortages. For instance, the New Zealand Government recently advised prospective university students to target areas of study and industry where there are long-term skill shortages (e.g. agriculture and forestry, engineering, education, health and social services, ICT, electronics and communication, and science). Similarly, the Canterbury region has special labour market needs because of the need to rebuild following recent earthquakes (e.g. engineering, surveying, project management, construction and urban planning).

Figure 36 The Christchurch rebuild has created a demand for skilled workers.

Figure 37 The Christchurch rebuild also generates economic activity in other sectors of the local economy.

Strategy 3: Increase accessibility to high-quality public services

Socio-economic position, ethnicity and gender are all related to significant inequalities among New Zealanders. Increasing accessibility to high-quality public services, such as education, health and family care, is a strategy that aims to overcome the inequalities that exist between different groups of New Zealanders.

The underlying causes of these inequalities are multi-faceted. There is, however, a strong correlation between people's geographic place of residence and their ability to access high-quality public services. For example, a family's location of residence affects their quality of health through:

- the accessibility the location provides to health services
- the accessibility the location provides to employment, educational opportunities and social services
- the availability of affordable, healthy food options
- factors such as the safety of the roads, recreational opportunities and public transport networks
- people's perceptions of their neighbourhoods and the degree of community cohesiveness
- the quality and appropriateness of the housing stock.

How each of these is addressed to counter inequality is often the subject of intense political debate. Strategies that have been suggested to help increase accessibility to high-quality public services include:

i Addressing operational issues, e.g. distribute health funds and resources according to need; the development of Maori and Pacific providers; the monitoring of health inequalities, social causes and the relationship between the two; and action policies that ensure equitable education, labour market, housing and other social outcomes.

ii Addressing pathways to access, e.g. housing initiatives; community development programmes; introduce programmes such as healthy cities and health-promoting schools; workplace interventions (such as Occupational Safety and Health); health education and the development of personal skills, and health protection.

iii Promoting health and disability services, e.g. improved access to appropriate, high-quality health care and disability services; implementation of the elective services booking system based on need; monitoring of service delivery to ensure equitable intervention rates according to ethnicity, gender, socio-economic status and region; primary care initiatives that reduce access barriers for Maori, Pacific peoples and other disadvantaged groups; ethnic-specific service delivery and equitable resource allocation by District Health Boards as funders and by providers.

iv Minimising the impact of disability and illness, e.g. income support (such as sickness benefits), disability allowance, accident compensation, anti-discrimination legislation and education, and support services for people with disabilities, chronic illness and mental health illness living in the community and their caregivers.

Learning Activities

1 Write a paragraph to explain how each of the following strategies could help reduce development inequalities in New Zealand:

   a   Reform tax and welfare policies

   b   Lower unemployment

   c   Increase accessibility to high-quality public services (e.g. health and education).

2 Suggest ways each of the strategies in Question 1 could be implemented.

3 As a class, discuss how New Zealand might change in economic well-being and 'quality of life' over the next 25 years.

4 Debate the costs and benefits of increasing taxes to redistribute income and reduce inequality.

Theme 2: India's place in the less developed world

India in profile

By global standards, India is a poor country. A third of the world's poor live in India and it has more people in poverty (i.e. living on less than the international benchmark level of $1.25 a day) than all of sub-Saharan Africa and China combined. Moreover, almost a fifth of the world's poor live in four Indian states alone: Andhra Pradesh, West Bengal, Orissa and Madhya Pradesh. In these states, tens of millions of children suffer from acute malnutrition, millions are illiterate, hundreds of thousands die of totally preventable diseases, and millions of children remain out of school.

Land area (sq km)		3,287,263
Population		1,210,193,422
Natural increase rate (%)		1.48
Infant mortality (per 1000 live births)		47.57
Life expectancy (years)	Male:	67.5
	Female:	72.6
Doctors per 1000 population		0.6
Literacy rate (%)	Male:	82
	Female:	65
GDP per capita (US$)		3,700
Gini coefficient		33.4
Population living in poverty (%)		25
Key industries	Textiles, chemicals, food processing, transportation equipment, cement, mining, petroleum, machinery, software, pharmaceuticals	
Labour force by sector (%)	Agriculture:	52
	Manufacturing:	14
	Services:	34
Land use (%)	Arable land:	48.3
	Permanent crops:	2.8
	Other:	48.4

Figure 38 India's statistical profile.

   ISBN: 9780170233316

It is for reasons such as this that India's Human Development Index (HDI) measure of 0.547 is much lower than that of New Zealand's (0.908), giving it a country rank of 136 out of 187 countries, a rank that is shared with Equatorial Guinea in sub-Saharan Africa.

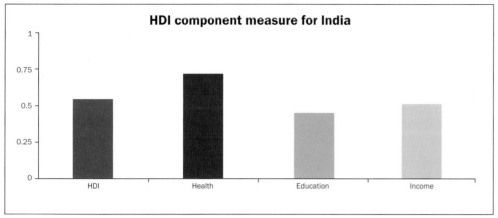

Figure 39 HDI components: health, education and income.

However, India's current HDI ranking fails to show just how much change the country has undergone in the last 30 years. A review of India's progress in each of the HDI indicators between 1980 and 2012 shows that India's life expectancy at birth increased by 10.5 years, mean years of schooling increased by 2.5 years and expected years of schooling increased by 4.4 years. The personal income of Indians has increased 273 percent between 1980 and 2012 (Figure 40).

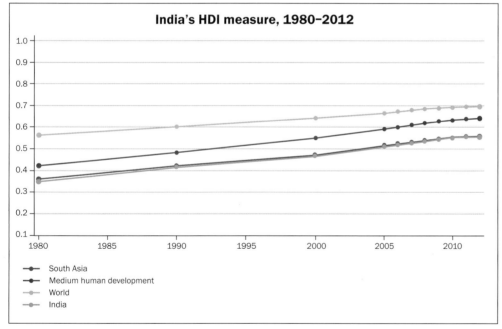

Figure 40 HDI trends, 1980 to present.

The three faces of India

India remains a country of extreme contrasts. Since being opened up in 1991, the Indian economy has grown rapidly. At a time when most Western economies are struggling to expand, India's continues to grow quickly, lifting millions out of poverty. This growth has also brought prosperity to many urban Indians. India now has 69 billionaires, has its own space programme, plans to send a man to the Moon, spends billions of dollars annually on defence and nuclear technology, and even has its own overseas aid programme.

Consequently, some suggest there are three 'faces' to India:
- 'Global India' (a reality for less than 20 percent of Indians)
- 'Developing India' (with some ties to 'Global India', and so seeing some benefits from the country's success)
- 'Poor India' (on the borderline between poverty and prosperity and as yet untouched by India's success).

Figure 41 'Global India'.

Figure 42 'Developing India'.

Figure 43 'Poor India'.

Overall, how does India's 'quality of life' compare with that of New Zealand's? Despite India's progress, if you lived in India instead of New Zealand you would most likely …

Have 10 times higher chance of dying in infancy:
- The number of deaths of infants under one year old in a given year per 1000 live births in India is 47.57 while in New Zealand it is 4.85.

Use 95% less electricity:
- The per capita consumption of electricity in India is 484 kWh while in New Zealand it is 9 228 kWh.

Consume 93% less oil:
- India consumes 0.0956 gallons of oil per day per capita while New Zealand consumes 1.5220.

Earn 89% less money:
- The GDP per capita in India is US$3,100 while in New Zealand it is US$27,300.

Die 14 years younger:
- The life expectancy at birth in India is 66.46 while in New Zealand it is 80.48.

Have twice as many babies:
- The annual number of births per 1000 people in India is 21.34 while in New Zealand it is 13.81.

Spend 95% less money on health care:
- Health spending per person in India is US$86 while New Zealand spends $2448.

Experience 1.66% more of a class divide:
- The Gini index measures the degree of inequality in the distribution of family income. In India it is 36.80 while in New Zealand it is 36.20.

Be three times more likely to have HIV/AIDS:
- The number of adults living with HIV/AIDS in India is 0.30% while in New Zealand it is 0.10%.

Have 47% more chance of being unemployed:
- India has an unemployment rate of 10.70% while New Zealand has 7.30%.

 ISBN: 9780170233316

Factors contributing to differences in India's development

Formerly a British colony, India gained independence in 1947. India's first decades of independence followed socialist-inspired policies and was characterised by state ownership, the extensive regulation of major industries and isolation from the world economy. Consequently, India's economy experienced slow economic growth for most of its independent history. It was not until the mid-1980s when India opened up its markets to global competition that its economy really started to grow. Unfortunately, the fruits of India's rapid economic growth have not always been shared evenly across its population, with the gap between the very rich and the very poor growing larger than ever before.

Factor 1: Population structure

In contrast to New Zealand's population pyramid (page 171), which has a narrower base and straighter sides, the shape of India's pyramid is broad at the base, indicating a high birth rate and a youthful population (Figure 44). The implication of India's large youthful population is that the youth will soon enter the workforce and also the reproductive age, resulting in demographic momentum (i.e. a situation where its population will continue to increase despite reduced reproductive rates).

In 2013, 30 percent of India's population was under the age 15, and only 8 percent were over the age of 65. According to the World Bank, India's youthful dependency ratio is high with 53 people (under 15 or over 65) for every 100 workers (aged 15–64).

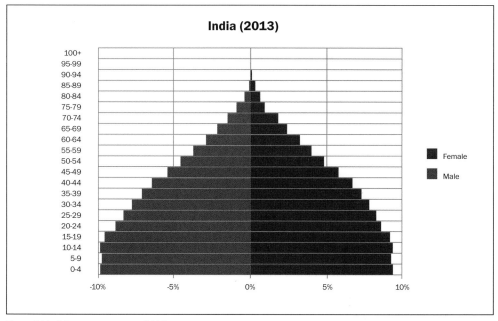

Figure 44 India's age-sex structure (2013).

ISBN: 9780170233316

Factor 2: Health

The World Bank estimates that India is ranked second in the world in terms of the number of children suffering from malnutrition, with 47 percent of the children exhibiting symptoms of malnutrition. One of the major causes of malnutrition in India is gender inequality. The diets of Indian women, due to their low social status, sometimes lacks in both quality and quantity. Furthermore, women who suffer malnutrition are more likely to have children that are underweight, undernourished and likely to die before the age of five.

At the same time, as a large number of India's population suffers from malnutrition, more than 100 million people (nearly 11 percent of India's population) are considered to be over-nourished. As a consequence, India has more people suffering from diabetes (50 million) than any other country in the world.

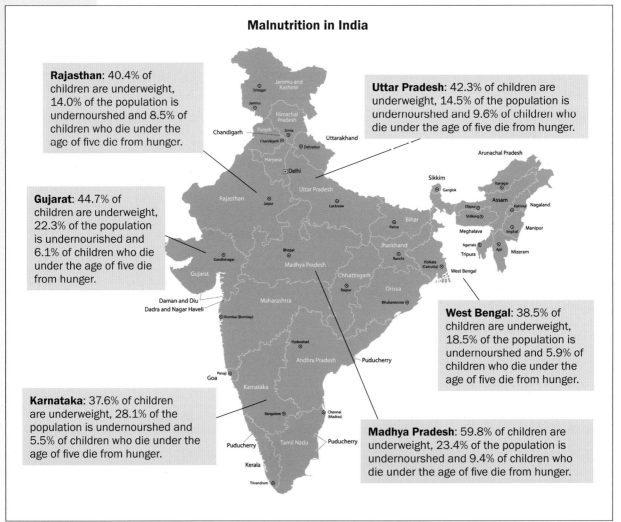

Figure 45

Factor 3: Housing

The quality of residential dwellings in India varies widely from modern apartment-style buildings to squatter settlements in big cities to tiny huts in rural villages.

India's largest city, Mumbai, with its population of 12.5 million, experiences wide disparities in housing quality between the affluent middle-income and low-income segments of the population. The highly desirable neighbourhoods Colaba, Malabar Hill, Marine Drive and Bandra accommodate professionals, those involved in the Bollywood movie industry and wealthy expatriates. Their apartments can have three or more bedrooms and are often outfitted with modern decor, furnishing and facilities. The Mumbai rich often inhabit gated communities and enjoy lush and peaceful gardens, and access to private pools and gyms.

However, these gated communities are thought to promote divisions (segregation) between ethnic, religious and racial groups in India. Social segregation between Hindus and Muslims is particularly widespread throughout India and sometimes results in communal ethic violence.

Yet despite Mumbai's recent economic growth, 60 percent of the city's residents still reside in informal housing or slums. India's largest slum, Dharavi, is located in central Mumbai and is now home to more than 1 million people. In rural areas, only 44 percent of houses have access to electricity, although this proportion rises in urban areas. Despite improved access to drinking water, no city in India provides a continuous water supply, and an estimated 400 millions Indians do not have access to a proper toilet.

Figure 46 Low-quality apartments.

Figure 47 Modern apartment living.

Figure 48 Slum housing.

Figure 49 Modern suburban living.

Factor 4: Economic standard of living

The Gini coefficient, or Gini index, measures the distribution or spread of income among individuals within a country deviates from a perfectly equal distribution. A Gini coefficient of 0 represents perfect equality (where everyone has the same income); a coefficient of 100 represents perfect inequality (where one person has all of the income and everybody else has zero income).

On a global scale, Japan, with a Gini coefficient of 24.9, has the lowest levels of inequality of all countries in terms of income distribution (Figure 50). Accordingly, Japanese society could be considered as egalitarian (equal), characterised by universal access to public goods and services, political stability and social cohesion. In stark contrast to Japan, the African nation of Botswana has the highest level of inequality, with a coefficient of 61.0. This suggests the wealth in Botswana is concentrated among a few to the exclusion of the majority. As a result, there is a high risk in Botswana of social unrest and civil conflict.

India has a Gini coefficient of 33.4 and New Zealand a coefficient of 36.2. This similarity suggests that both New Zealand and India have comparable levels of inequality.

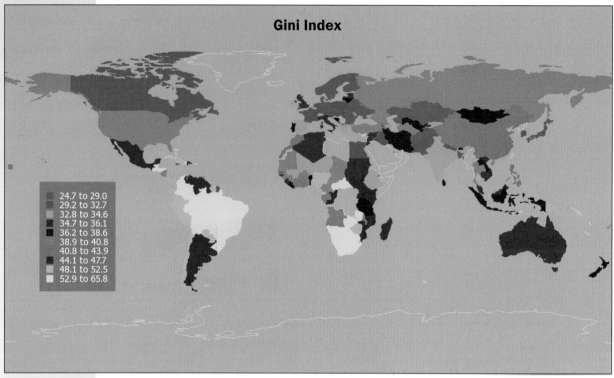

Gini Index

24.7 to 29.0
29.2 to 32.7
32.8 to 34.6
34.7 to 36.1
36.2 to 38.6
38.9 to 40.8
40.8 to 43.9
44.1 to 47.7
48.1 to 52.5
52.9 to 65.8

Figure 50 Gini measures for selected countries, 2012.

Figure 51 Slum dweller child.

Figure 52 Homelessness.

Factor 5: Gender discrimination

Despite recent improvements in female literacy and life expectancy, India's last census in 2011 counted only 914 girls aged 6 and under for every 1000 boys (Figure 48). Without human intervention there should be marginally more boys born into a population than girls. Consequently, had a normal sex ratio prevailed, India would have an additional 600,000 baby girls a year added to the population than is currently the case. This proposes that in India, gender discrimination begins before birth as a result of pre-natal gender selection.

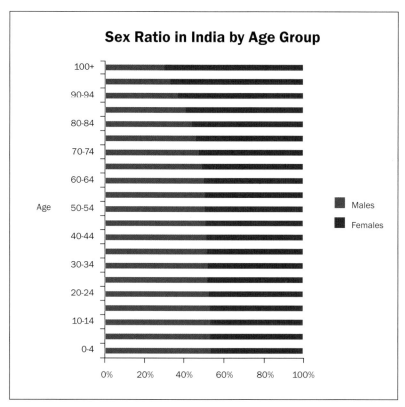

Figure 53 Male-female gender ratios.

Evidence suggests that India's preference for sons is not caused by poverty or a lack of education, as the states with the worst sex ratios (Punjab, Haryana and Gujarat) are also among the richest. Rather, parents choose to abort female foetuses because they feel for social reasons they must have sons, particularly if they want to keep the family small. To address this, India has introduced laws to ban ultrasound scans from being used merely to identify a foetus's sex, and sex-selective abortions are illegal. However, what is really needed is a change in social attitude to persuade parents that daughters are worth as much as their sons. Although it is not easy for governments to change social attitudes, they can help by ensuring that girls get their fair share of education, and women their fair share of health care.

Nevertheless, if sex ratios stay the same, 600,000 missing girls this year will become, in 18 years' time, more than 20 million missing future brides. History shows that any society with large groups of young single men is likely to experience increases in robbery, rape and bride trafficking.

Factor 6: Civil and political rights

Aimed at overturning the inequalities of past social practices, the civil and political rights of Indian citizens are outlined in a document called the Constitution of India. The constitution guarantees civil liberties common to most democracies, such as:

- equality before law
- freedom of speech and expression
- peaceful assembly
- the freedom to practise religion.

Specifically, it aims to prohibit any discrimination on the grounds of religion, race, caste, sex or place of birth. In reality, though, those of the traditional lower castes (untouchables) remain marginalised in Indian society. For these people, who account for 16.5 percent (or 170 million) of India's population, matters of child labour and child marriage, and oppressive social practices like sati (the wife following her dead husband onto the funeral pyre) remain real issues.

Factor 7: Knowledge and skills

In India, about 12 million people join the workforce each year comprising highly skilled, skilled, semi-skilled and unskilled workers. The last category constitutes the majority of the population entering the workforce. Rapid economic growth in India has increased the demand for highly skilled and semi-skilled labour, resulting in a shortage of skilled workers in the country. As a result, India is among the top countries in which employers are facing difficulties in filling up the jobs.

The training of the school-leaving population is a problem that needs to be addressed if India's fast-paced economic growth is to continue. Unlike richer countries where there is a burden of an ageing population, India has a unique 20–25-year window of opportunity called the 'demographic dividend'. This 'demographic dividend' means that for a limited period of time, India will have a higher proportion of a working-age population compared with other countries relative to its entire population.

Figure 54 An examination session in Jaura, India.

Factor 8: Social cohesion

India's caste system has influenced aspects of Indian society for many centuries. Initially rooted in the Hindu religion and based on a division of labour, the caste system dictates the kind of occupations an individual can pursue and the social interactions they can have. Similar to the social class structure found in other societies, castes are ranked in hierarchical order and determine how the people of one caste are to relate to that of another.

India's caste system has five main classes. In descending order, the castes are as follows:

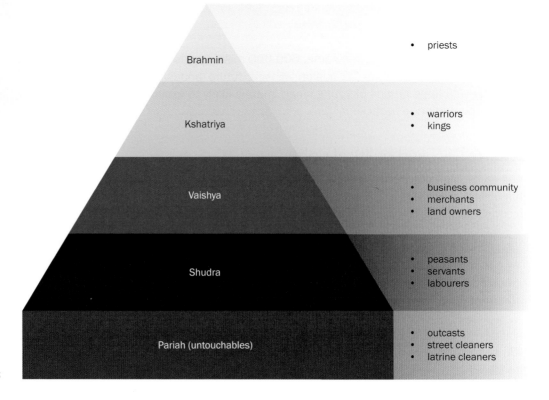

Figure 55

Because of its rigidity, the caste system is contrary to any form of social cohesion. This is because:
- The family an individual is born into determines one's caste for life.
- Castes rarely intermarry and are not changeable.
- Lower castes are prevented from aspiring to higher castes, limiting the economic progress of the poorest.

Since gaining independence from Britain in 1947, India has introduced many laws and social initiatives to protect and improve the socio-economic conditions of its lower castes, however despite improvement in the treatment of the lower castes in some areas, progress remains slow in others.

Factor 9: Natural hazards

Figure 56 A Brahmin.

India's climate and geographical location places it at high risk from natural hazards. Moreover, India's vulnerability to hazards is increasing over time due to the increased concentration of people in urban centres, environmental degradation and a lack of planning and preparedness.

The highest hazard risk is from its susceptibility to flash floods caused by unpredictable and heavy monsoonal rain. In recent years, floods have killed thousands and displaced millions but the greatest impact of floods has been on poor rural communities whose crops, which are their only source of income, are frequently destroyed by flood events (Figures 57 and 58).

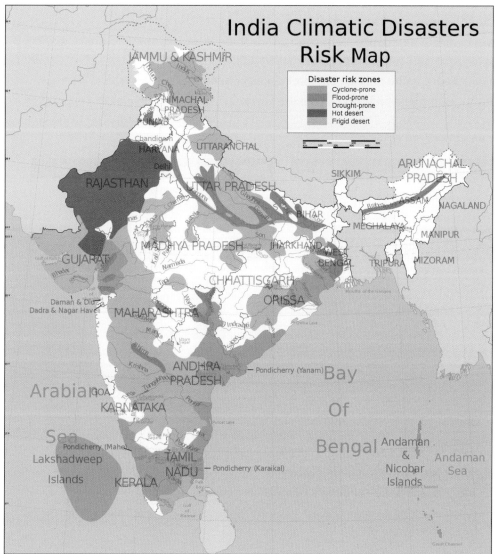

Figure 57 Natural hazards map of India.

Other threats posed to India's population by the natural environment include droughts, hurricanes, earthquakes and landslides (Figure 58). Most notable was tropical cyclone Orissa, which in 1999 brought storm surges and heavy rains, killed 9803 people and left 2 million homeless.

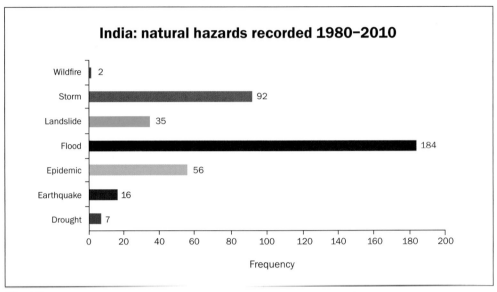

Figure 58

Learning Activities

1   Why is India considered a 'less developed' country? Support your answer with quantitative evidence from the text.

2   Using qualitative evidence from the text, describe how the 'quality of life' in India differs from that experienced by New Zealanders.

3   What is meant by the phrase 'the three faces of India'?

4   Using the data in the table and an outline regional map of India, construct a choropleth map to show regional disparities in development. Write a brief paragraph describing the pattern shown in the map.

Rank	Indian state	HDI
1	Kerala	0.790
2	Delhi	0.750
Medium human development		
3	Himachal Pradesh	0.652
4	Goa	0.617
5	Punjab	0.605
6	Northeastern India (excluding Assam)	0.573
7	Maharashtra	0.572
8	Tamil Nadu	0.570
9	Haryana	0.552
10	Jammu and Kashmir	0.529
11	Gujarat	0.527
12	Karnataka	0.519

Rank	Indian state	HDI
Low human development		
13	West Bengal	0.492
14	Uttarakhand	0.490
15	Andhra Pradesh	0.473
	India (national average)	0.467
16	Assam	0.444
17	Rajasthan	0.434
18	Uttar Pradesh	0.380
19	Jharkhand	0.376
20	Madhya Pradesh	0.375
21	Bihar	0.367
22	Odessa	0.362
23	Chhattisgarh	0.358

5   Study the population pyramid in Figure 44.

   a   Describe the shape of India's population pyramid.

   b   What challenges does India's youthful population present with regard to its future development?

6 Describe the main characteristics of housing shown on page 187.

7 Using the data in the table, construct a scatter graph to illustrate regional differences in home ownership and household size. Write a statement describing the pattern shown in the graph.

States	Families owning houses (%)	Household size
Andhra Pradesh	56.0	3.9
Arunachal Pradesh	20.4	4.8
Assam	19.7	4.7
Bihar	20.1	5.4
Chhattisgarh	17.2	5.0
Gujarat	73.2	4.2
Haryana	78.5	4.7
Himachal Pradesh	52.2	5.3
Jammu and Kashmir	50.2	5.7
Jharkhand	28.1	5.4
Karnataka	54.6	4.6
Kerala	87.5	4.3
Madhya Pradesh	24.8	5.0

States	Families owning houses (%)	Household size
Maharashtra	67.2	4.7
Manipur	24.8	5.0
Meghalaya	34.9	5.1
Mizoram	22.8	4.8
Nagaland	20.6	4.5
Odessa	31.7	4.5
Punjab	77.2	5.2
Rajasthan	40.4	5.5
Sikkim	50.9	4.5
Tamilnadu	58.5	3.5
Tripura	11.9	4.3
Uttar Pradesh	27.3	5.7
Uttarakhand	48.8	5.0
West Bengal	60.1	4.5

8 Study Figure 53.

    a Describe the differences in the male-female ratio from birth to old age.

    b What are the implications of an unbalanced male-female sex ratio?

9 Study the natural hazards map of India in Figure 57 and Figure 58.

    a Identify the hazard that is most widespread in India.

    b Which regions of India are at greatest risk from natural hazards?

    c Describe the distribution of flood-prone areas.

10 Rank the following factors that have contributed towards India's level of development from most influential to least influential. Write a paragraph justifying your ranking.

Population structure	Health	Quality of housing	Caste system
Natural environment	Religion	Civil and political rights	Education and skills
Built environment	Gender inequality	Economic standard of living	History

Strategies for reducing development inequalities

Strategy 1: Trade and foreign direct investment

Trade plays an important role in the development and global distribution of wealth by enabling the labour markets of less developed countries to access those of more developed countries. The benefit of free trade as a strategy for reducing differences in development is that it enables many developing countries to develop competitive advantages in the manufacture of certain products. Free trade frequently benefits the poor because the increased economic growth that results from freer trade itself tends to increase the incomes of the poor as well as those of the population as a whole, and new jobs are created for unskilled workers, raising them into the middle classes.

Figure 59 Indian migrant workers in Dubai.

Figure 60

For some developing countries, the international trade of manufactured goods has come about as a result of foreign direct investment (FDI) in specialised industries. For example, India is considered the second most favoured FDI destination (after China) for transnational corporations (TNCs). The sectors that attract high inflows of FDI in India are services, telecommunication, construction activities and the production of computer software and hardware.

Examples of TNCs recently investing in India include the following:

- Dashtag, a UK-based drug company, has plans for the development of pharmaceuticals specialising in dermatology, antihistamines, antibiotics and oncology products.
- Japanese automaker Nissan intends to produce 10 new passenger car models in India by 2016 in a bid to boost its sales volumes in India and eventually gain 8 percent of the Indian car market by 2016.
- Italian agricultural machinery producer Maschio Gaspardo Group has ventured into India by establishing a new facility near Pune (India's seventh-largest city) at a cost of US$37 million. The company plans to invest an additional US$18.86 million over 2012–17. With an annual capacity of 20,000 units, the Pune plant would initially manufacture rotary tillers, mulchers and seeders for the Indian market.

However, trade barriers and tariffs often limit the trade of goods and services between countries. A regional trading bloc is a group of countries within a geographical region that engage in international trade together, and are usually related through a free trade agreement. Trading blocs are a form of economic integration, and shape the pattern of world trade (Figure 35). Non-member countries that wish to trade with countries that are members of a training bloc are often subjected to trade restrictions such as tariffs, which are extra taxes imposed on goods imported or exported from another country.

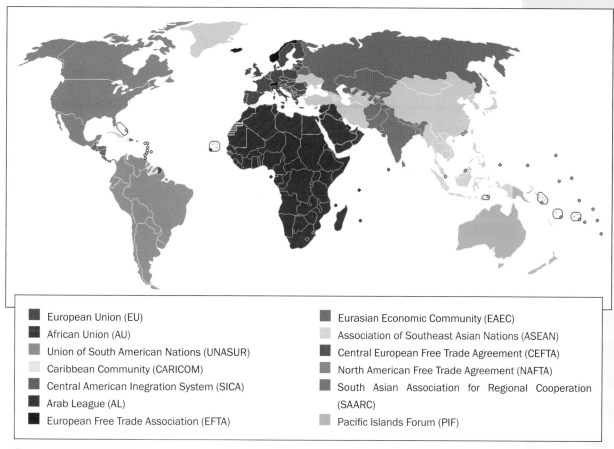

European Union (EU)
African Union (AU)
Union of South American Nations (UNASUR)
Caribbean Community (CARICOM)
Central American Inegration System (SICA)
Arab League (AL)
European Free Trade Association (EFTA)

Eurasian Economic Community (EAEC)
Association of Southeast Asian Nations (ASEAN)
Central European Free Trade Agreement (CEFTA)
North American Free Trade Agreement (NAFTA)
South Asian Association for Regional Cooperation (SAARC)
Pacific Islands Forum (PIF)

Figure 61 Regional trade blocs.

The World Bank estimates that the removal of global trade barriers could lift as many as 500 million people out of poverty globally and inject US$200 billion annually into the economies of developing countries. China, for example, a country that has aggressively opened its markets and expanded its trade, saw poverty decline by 377 million. India has an economic cooperation agreement with neighbouring countries Bangladesh, Bhutan, Maldives, Nepal, Pakistan and Sri Lanka (Figure 58) called the South Asian Association for Regional Cooperation, or SAARC. One of the main goals of SAARC is to work towards the removal of all barriers to trade between member nations with the objective of creating a free trade zone in South Asia.

New Zealand also is a signatory to several free trade agreements. However, its most important is the bilateral agreement it has with Australia, its largest export partner. Known as the Closer Economic Relationship (CER), the agreement between New Zealand and Australia aims to support the free flow of merchandise trade, services, investment, labour and visitors between the two countries.

An alternate strategy to free trade is *fair trade*. Fair trade is trade in which fair prices are paid to producers in developing countries. According to the World Trade Organisation (WTO), the concept of fair trade is 'a trading partnership, based on dialogue, transparency and respect, that seeks greater equity in international trade' and encourages sustainable development by offering 'better trading conditions to, and securing the rights of, marginalized producers and workers'. India is the principal exporter of fair trade cotton.

Figure 62

Strategy 2: Remittances

A remittance is a transfer of money by a foreign worker to his or her country of origin. Remittances account for the second largest inflow of money to developing countries and exceed by a factor of three that which is transferred to less developed countries in the form of international aid. Remittances play an important role in the global transfer of wealth. They contribute to the economy of developing countries and to the livelihoods of the poor.

Remittance flows to the less developed world exceeded US$406 billion in 2012 and are projected to grow to US$685 billion in 2015. The top recipient of officially recorded remittances in 2012 was India, which received US$70 billion, followed by China with US$66 billion. Other large recipients include the Philippines, Mexico and Nigeria (Figure 63).

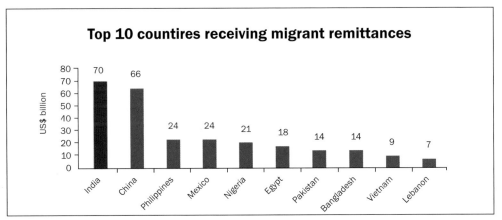

Figure 63 Top 10 countries receiving migrant remittances (US$ billion, 2012).

Figure 64

For some countries, remittances make up a significant proportion of the country's total earnings (gross domestic product, or GDP). Remittances sent by foreign Tajik workers accounted for 47 percent of Tajikistan's GDP in 2011. In the same year, remittances sent to Samoa accounted for 21 percent of the country's total GDP.

Strategy 3: International aid

The effectiveness of aid as a tool for development has been the topic of intense debate. Although aid assistance will never match that of direct investment by trans-national corporations (TNCs), many poor countries seek foreign aid in the form of loans and grants because:

- they lack the hard currency to pay for foreign goods such as oil and machinery required for development
- population pressure places a strain on local expenditure, leaving little capital (money) to invest in industry and infrastructure
- the local population lack the 'know-how' or skills required for development.

Traditional sources of aid usually fall into one of two categories (Figure 65):

i Official government aid refers to that which is given by the government of a foreign country. Official government aid can be divided into two types: bilateral aid is given directly from one country to another; multilateral aid is given by a group of countries usually under the umbrella of an international body such as the United Nations.

ii Voluntary aid is organised by non-governmental organisations (NGOs) or charities. Examples of such organisations include UNICEF, World Vision and Oxfam.

Aid is supplied to countries as either short-term emergency aid, which is provided to help a country or region to cope with unexpected disasters such as drought, flooding and tropical cyclones, or long-term development aid, which is used to fund initiatives that promote continuous improvement in the quality of life of communities in a poorer country.

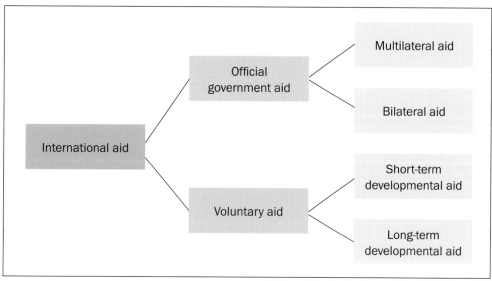

Figure 65 Types of international aid.

Critics of traditional forms of aid argue that the 'top-down' approach to providing aid can create a culture of dependency and that too often the benefits of the funding given are either short-lived or fail to reach the very poorest people in a community.

An alternative approach is to give small amounts of aid directly to those who most need it. Known as the 'bottom-up' approach, this form of aid assistance seeks to empower the impoverished by giving them start-up loans (micro-finance) and equipping them with the necessary skills to enable them to improve their own quality of life and that of their communities (Figure 66).

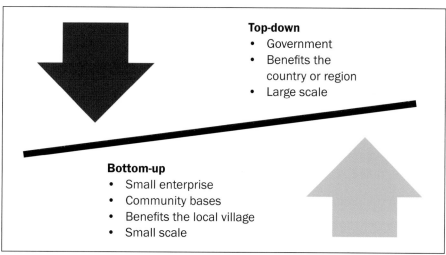

Figure 66 Top-down versus bottom-up aid strategies.

ISBN: 9780170233316

Founded in 2005, the charity organisation Kiva is an example of a non-profit organisation that uses a 'bottom-up' approach to delivering aid to reduce poverty in India's poorest communities. Using the Internet and a worldwide network of micro-finance institutions, Kiva lets individuals lend as little as $25 to people around the world without access to traditional banking systems. In doing so it provides affordable access to capital (money) to those in need, which in turn helps people create better lives for themselves and their families.

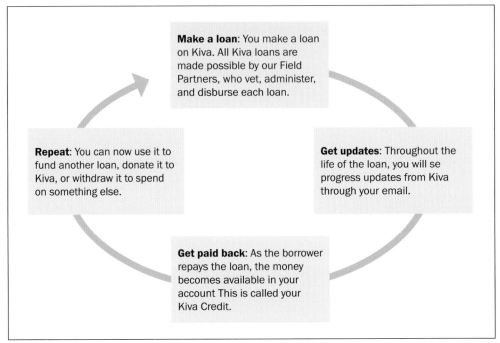

Make a loan: You make a loan on Kiva. All Kiva loans are made possible by our Field Partners, who vet, administer, and disburse each loan.

Get updates: Throughout the life of the loan, you will se progress updates from Kiva through your email.

Get paid back: As the borrower repays the loan, the money becomes available in your account This is called your Kiva Credit.

Repeat: You can now use it to fund another loan, donate it to Kiva, or withdraw it to spend on something else.

Figure 67 Kiva's 'bottom-up' approach to aid.

Figure 68

In rural India, millions of people (mostly women) live without access to finance and as a result struggle to make ends meet. Kiva seeks to aid the rural poor by working with individual families, discussing their financial needs, and then by sourcing funding in the form of micro-finance from willing lenders. Micro-finance is a term used to describe financial services to low-income individuals or to those who do not have access to typical banking services. The idea behind micro-finance is the belief that the poor are capable of lifting themselves out of poverty if given access to financial services. Accordingly, micro-finance loans have been used in rural India in recent years to support a number of initiatives. In particular, Kiva has given support to developments that:

• support excluded and vulnerable groups like widowed women, physically challenged individuals and families affected by leprosy
• improve drinking water and sanitation systems
• promote solar and renewable sources of energy.

 ISBN: 9780170233316

'We were struggling to meet our daily needs. We had to borrow money. Life was hard,' said Sumita Das, 24, about her life a few years ago.

Sumita, her husband Champak Das, and their son subsisted on her husband's weekly wage of Rs.300. Sumita earned another Rs.200 by weaving traditional Assamese clothing like mekhala chadar and gamchas woven on a single loom in their cramped one-room house.

'Then, I took a loan from Bandhan (a microfinance company), bought more looms and raw material. My husband began to work from home. We now have three looms and one of our neighbors works for us,' she said.

'Now, we earn Rs.9,000 a month. We have a colour television, a home theatre, a fan, a mobile phone and a power inverter. We also send our only son to a private school. After paying a monthly instalment of Rs.1,680, we save Rs.3,000 every month. All this was unthinkable for us a few years back,' said Sumita, a smile breaking out on her lips.

Sumita Das, Kaibortapara, Sualkuchi, Assam

Case Study 1

Poor India: UNDP-IKEA partnership helps women change rules in India

For most of her life, Shiela Devi had few options. Growing up in an impoverished village in India's northern state of Uttar Pradesh, she ate her brothers' leftovers, did chores at home while they went to school, and married at age 15.

At 35, Shiela was completely illiterate, had no source of income, and could not imagine any other way to raise her young daughter. Shiela's life was not much different from the lives of many women in Uttar Pradesh, where over 30% of the population lives below the poverty line and less than half of the women are literate. Shiela certainly never dreamed she could go into business for herself, generate a profit, or influence decisions within her household and community.

Yet, thanks to a joint United Nations Development Programme (UNDP)-IKEA partnership, women like Shiela are not only dreaming, but they are also doing. Launched in 2009, the five-year programme has created 238 self-help groups to boost literacy and leadership among 50,000 women in the 500 villages of the districts of Jaunpur, Mirzapur and Sant Ravidas Nagar.

Through these networks, women gain financial literacy and are educated on many other issues as well, including domestic violence, legal aid, their rights to information and property inheritance, and child labour.

It was through such a group that Shiela attended a discussion about nutrition and the practice of feeding girls last. After learning about the dangers of this practice, she is now determined that her daughter will be well fed, as well as educated. The groups have also created new awareness among women of the importance of participating in local decision-making and democratic processes, such as council elections.

Case Study 1

To affirm their collective strength, these women have signed a 12-point charter that spells out what empowerment means to them. Travelling through the 500 villages, the 10-foot tall charter serves as a powerful reminder to women that they are not alone and can change the rules. 'When the charter was being drawn up I suggested that girls should not be married before the age of 18,' says 40-year-old Susheela Devi who was married at 13.

Decades of poverty, deep-rooted caste hierarchies, and gender inequality do not change overnight. But in Uttar Pradesh, thousands of women are working together to empower themselves and create new opportunities for social change.

Figure 69

Figure 70

Strategy 4: Debt relief

The burden of debt is considered as one of the main barriers that prevent poorer countries from developing. This is because countries already heavily indebted are often forced to raise additional debt to meet the interest and principal repayments of current debt, trapping the country into a continuous cycle of ever-increasing debt.

The International Monetary Fund (IMF) and the World Bank established the Heavily Indebted Poor Countries (HIPC) initiative in 1996 with the aim of ensuring that no poor country faces a debt burden it cannot manage and to provide a comprehensive strategy to reduce the debt of the world's most heavily indebted countries. To qualify for HIPC assistance, countries had to demonstrate:

- commitment to an IMF programme and progress in developing a national poverty strategy
- reforms and sound policies for economic growth, human development and poverty reduction.

Of the 39 countries identified by the HIPC as heavily indebted since its inception, 33 of are in sub-Saharan Africa.

Strategy 5: Adherence to the Millennium Development Goals

The Millennium Development Goals (MDGs) are eight international development goals agreed to by the 193 member countries of the United Nations in September 2000, as a global strategy to reduce or eradicate extreme poverty and its effects.

The goals, which range from eradicating extreme poverty to stopping the spread of HIV/AIDS and providing primary education for all by the target date of 2015, are mostly targeted towards developing countries, but they also include roles for developed countries like New Zealand. For example, Goal 8 sets objectives for developed countries to enhance debt relief for the least developed countries and make technological benefits (such as telephones, computers and access to the Internet) available to them.

The eight goals are:

ERADICATE EXTREME POVERTY AND HUNGER

ACHIEVE UNIVERSAL PRIMARY EDUCATION

PROMOTE GENDER EQUALITY AND EMPOWER WOMEN

REDUCE CHILD MORTALITY

IMPROVE MATERNAL HEALTH

COMBAT HIV/AIDS, MALARIA AND OTHER DISEASES

ENSURE ENVIRONMENTAL SUSTAINABILITY

A GLOBAL PARTNERSHIP FOR DEVELOPMENT

India's progress towards achieving the MDGs by the 2015 target date has been tracked against 33 targets and 60 indicators addressing issues such as extreme poverty and hunger, education, women's empowerment and gender equality, health, environmental sustainability and global partnership. Thus far, India has made encouraging progress towards most of the MDGs and given India is one of the most populous countries in the world, success in India will affect the world's ability to achieve the MDGs.

In summary, India had some successes and some failures. Notable successes include:

- The percentage of the population in poverty (defined as the number living on less than US$1.25 a day) has declined from 45 percent in 1993 to 37 percent in 2005 (Goal 1).

- India has achieved the 95 percent cut-off line regarded as the marker value for achieving the 2015 target of universal primary education for all children aged 6–10 years (Goal 2).

- India is close to attaining 100 percent youth literacy rate of 15–24 years olds (Goal 2).

- The gender disparity in all the three grades of education (primary, secondary and tertiary levels) has been steadily diminishing over the years (Goal 3).

- The under-five mortality rate (expressed as a rate per 1000 live births) has declined from 117 (per 1000) during the last decade but still varies from 71 in rural areas to 41 in urban areas (Goal 4).

- In the last 20 years, infant mortality (defined as the deaths of infants of age less than one year per thousand live births) has declined by an average of 1.5 points per year over the last 20 years (Goal 4).

- The number of women who die from any cause related to pregnancy has declined from 327 deaths per 100,000 births in 2001 to 139 deaths per 100,000 births in 2009 (Goal 5), however only 47 percent of births are attended by health care professionals.

- Among pregnant women of 15–24 years, the prevalence of HIV has declined from 0.86 percent in 2004 to 0.48 percent in 2008 (Goal 6).

- A network of 668 protected areas has been established, extending over 1,61,221.57 sq km (4.90 percent of total geographic, comprising 102 national parks, 515 wildlife sanctuaries and 47 conservation reserves). Furthermore, 39 tiger reserves and 28 elephant reserves have been designated for species-specific habitat management (Goal 7).

- Over a period of 12 years, India's Internet subscriber base has increased 97-fold from 0.21 million in 1999 to 20.33 million in 2011 (Goal 8).

However, notable failures include:

- India has made slow progress in eliminating the effect of malnourishment in children below three years of age (Goal 1).

- Labour markets in industry and service sectors in India remain male dominated. The percentage share of females in paid employment in the non-agricultural sector is extremely low at 18.6 percent (Goal 3).

- India is likely to fall short of universal immunisation of one year olds against measles (Goal 4).

- Energy consumption per capita is increasing at an average rate of 11 percent per year (Goal 7).

- The proportion of households without improved sanitation facilities is still far behind that which is required. Estimates suggest that as few as 42 percent of households have access to a sanitation facility, which is less than those who have access to a mobile phone.

Figure 71

 ISBN: 9780170233316

The table summarises India's progress towards achieving the MDGs so far.

Goal 1: Eradicate extreme poverty and hunger		1993-94	2004-05
1	Poverty Headcount ratio (%)	45	37
2	Poverty Gap Ratio (rural)	8.5	5.7
3	Poverty Gap Ratio (urban)	8.1	6.1
4	Share of poorest quintile in National Consumption (rural) (%)	9.6	9.5
5	Share of poorest quintile in National Consumption (urban) (%)	8	7.3
		1998-99	2005-06
6	Prevalence of underweight children under 3 years of age (%)	42.7	40.4

Goal 2: Achieve universal primary education		2007-08	2009-10
7	Net Enrolment Ratio (%)	95.9	98.3
		1999	2008-09
8	Proportions of pupils starting Grade 1 who reach Grade 5 (%)	62	76
		2001	2007-08
9	Literacy rate of 15-24 years olds (male) (%)	76.7	91
10	Literacy rate of 15-24 years olds (female) (%)	54.9	80

Goal 3: Promote gender equality and empower women		2004-05	2007-08
11	Gender Parity Index (Primary)	0.95	0.98
12	Gender Parity Index (Secondary)	0.79	0.85
13	Gender Parity Index (Tertiary)	0.71	0.7
		2001	2007-08
14	Ratio of literate women to men, 15-24 years old	0.8	0.88
		1999	2011
15	Proportion of seats held by women in parliament (%)	9.6	10.96

Goal 4: Reduce child mortality		2005	2009
16	U5MR (per 1000)	74.3	64
		2006	2009
17	IMR (per 1000 live births)	57	50
		1993	2009
18	Proportion of one year olds immunised against measles (%)	42.2	72.4

ISBN: 9780170233316

Goal 5: Improve maternal health		2004-06	2007-09
19	Maternal Mortality Ratio (MMR) (per 1,000,000 live births)	254	212
		1992-93	2007-08
20	Proportion of births attended by skilled health professionals (%)	33	52

Goal 6: Combat HIV/AIDS, malaria and other diseases		2004	2008
21	HIV prevalence among pregnant women aged 15-24 years old (%)	0.86	0.48
		2006	2010
22	Incidence rate of malaria (%)	1.67	1.47
23	Deaths per 100 due to malaria	0.1	0.06
		2008	2010
24	Prevalence of TB per 100,000 population	248	256
25	Mortality due to TB per 100,000 population	23	26

Goal 7: Ensure environmental sustainability		2003	2007
26	Proportion of land area covered by forest (%)	20.64	21.02
		2009	2011
27	Ration of area protected to maintain biological diversity to surface area	4.83	4.9
		2007	2009
28	CO_2 emissions per capita (metric tons)	1.21	1.37
		2003	2007
29	Consumption of ozone depleting CFC (ODP tonnes)	2608	998.5
		2005-06	2008-09
30	Proportion of population with sustainable access to an improved water source (%)	88	91.4
		2005-06	2007-08
31	Proportion of population with access to improved sanitation (%)	40.6	42.3
			2001
32	Slum population as percentage of urban population		23.1

Goal 8: Develop a global partnership for development		2009	2011
33	Telephone lines and cellular subscribers per 100 population	44.88	73.97

Figure 69 India's progress towards achieving the MDGs.

Learning Activities

1 As a class, discuss how India might change in economic well-being and quality of life over the next 25 years.

2 Outline the advantages and disadvantages of trading agreements.

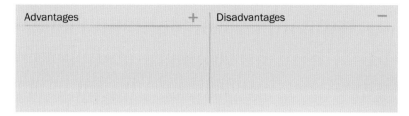

Advantages	+	Disadvantages	−

3 How do remittances help India?

4 Explain why bottom-up forms of aid are more effective than top-down forms of aid.

5 Write a paragraph to explain how each of the following strategies could help reduce inequalities in India:

a freer trade

b encourage more foreign direct investment

c increase remittances receipts

d international aid

e debt relief

f a commitment to the MDGs.

6 Suggest ways that each of the strategies in Question 5 could be implemented.

7 With reference to Figure 69, summarise India's progress towards each of the MDGs thus far. Is India on target to meet all of the MDGs by 2015?

8 Which, in your opinion, are the three most important MDGs?

9 Go to the United Nations MDGs website (www.mdgmonitor.org or www.un.org/millenniumgoals) and as a group investigate the global progress made so far in achieving each of the MDGs.

Contemporary
Geographic
Link to

New Zealand Issue: A Rail Auckland Airport?

5 Planes, trains and automobiles: A Rail Link to Auckland Airport?

A contemporary New Zealand geographic issue

There have been proposals to provide a railway connection to Auckland Airport. This study investigates the issue: 'Should such an airport rail connection be built?'

Figure 1

Figure 2

Airports and geography

'An airport is the place that gives people their first ... and last ... impression of a city and a country. If an airport works it becomes an easy and pleasurable place for people to visit for pleasure or business. And when it doesn't some may never come back.'

— Wendy Waters, *All About Cities*

- Airports are places of movement and connection for people and for cargo. They are origin and destination gateways.
- Airports are all about accessibility. They must be accessible and provide easy movement in and out for people and cargo. This accessibility is provided by air and land connections with other places.
- Airports are hub or nodal places: they are central points in a system or network of connections between places (Figure 3).

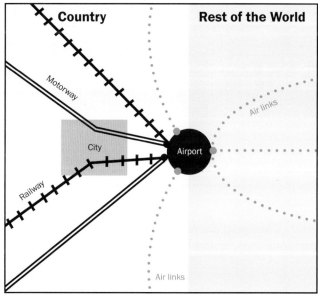

Figure 3 Airports are central points in transport networks.

Aerotropolis, or airport city

Air transportation moves more people and goods faster and further than ever before. As a result, new urban developments are taking place around airports that form a cluster of activities related to passenger and cargo flows: offices, hotels, convention centres, entertainment, freight storage and distribution centres, industrial buildings, shopping centres and residential and retail districts.

A whole city or part of a city based on and around an airport and air travel is known as an aerotropolis (Figure 4). These may be either brand new cities or old cities that are transforming and renewing. Transport is the heart of the aerotropolis: air, road and rail links provide the connections for the fast and efficient flow of passengers and freight between the aerotropolis and other places in the home country and the rest of the world.

Examples of cities and places with developed and growing aerotropolises are Amsterdam and London in Europe, Dallas/Fort Worth and Chicago in the USA, Singapore and Beijing in Asia, Dubai and Abu Dhabi in the Middle East. Most developments are at present located in wealthier countries. Brisbane is the only recognised example from the Oceania region. Auckland, with the growing business and commercial centres located close to the airport, could soon join this list.

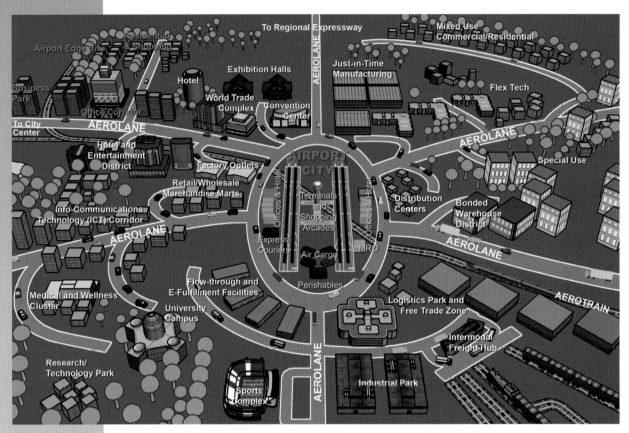

Figure 4 *The aerotropolis, or airport city.*

In the 21st century the airport has become the new focal growth point of many cities, replacing focal points of the past (Figure 5).

Changing drivers of urban development, economic growth and business location within the city

Date	Driver and focus
18th century	Seaports and ships
19th century	Railways and trains
20th century	Roads and cars
21st century	Airports and planes

Figure 5

Geographers have an interest in studying airports because airports are about:

- Location
- Movements and connections between places
- Movements and connections between people
- Accessibility
- Growth and change.

Figure 6 *Airport geography.*

1   Copy the Wendy Waters quote from page 209. Make a bullet point list of things about an airport that would create a favourable impression for visitors. Aim for a list of five points.

2   Figure 3 illustrates that airports are nodal places. Draw a diagram by adapting Figure 3 to illustrate one of these other two concepts:

Either **a** Airports are places of movement and connection for people and for cargo. They are origin and destination gateways.

Or **b** Airports are all about accessibility. They must be accessible and provide easy movement in and out for people and cargo. This accessibility is provided by air and land connections with other places.

3   Refer to the text and Figure 4.

Either **a** Draw a simplified diagram of an aerotropolis. Emphasise three things:

i   the airport itself;

ii   road and rail transport features;

iii services and facilities around the airport.

Or **b i** List five features of an aerotropolis, and **ii** explain why airports become focal places for growth and development. Refer to concepts like accessibility, location, linked activities and cumulative causation in your answer.

4   Refer to Figure 5. How and why do the drivers of urban growth and economic development change over time?

Auckland Airport

The contemporary issue is: 'Should a rail connection to Auckland Airport be built?'

Figure 7 Auckland Airport.

In the news

Auckland's $2.4 billion airport of the future

Artist's impression of Auckland Airport's 30-year vision for a new domestic and international terminal

Auckland Airport Company has released plans for 'the airport of the future' that includes a sweeping crescent-shaped domestic and international terminal on one site.

The project, which has a price tag of $2.4 billion, is part of the company's strategy of building the airport's position as a key hub in the Asia-Pacific region.

Forecasts suggest the number of passenger movements through the airport will increase from around 14.5 million today to 40 million by 2044 and the airport says it needs to plan for sufficient infrastructure to cope with that growth.

In its presentation the company said investment was expected to be staged to provide a reasonable price path and a fair return. The company said that passengers would end up paying for the development but analysis suggested increase in charges would be in line with rises in the consumer price index and that 'people accept if they want a great airport facility there is a cost'.

The company says it does not expect to have to build a second runway until around 2025. In 2009 the company put the brakes on a second runway development because of a drop-off in the rate of aircraft movement growth but passenger growth was increasing at a higher rate, mainly because of bigger planes.

The plan stated there was a 'rare opportunity' to create a uniquely New Zealand airport environment including a stand of native trees and plants at the entrance of the building. The plan allows for new or extended aircraft piers with provision for new hotels and commercial space next door. The airport is also allowing for a train terminal although it says any rail link would have to be built by central or local government.

Figure 8 Auckland Airport — looking to the future.

Location of Auckland Airport

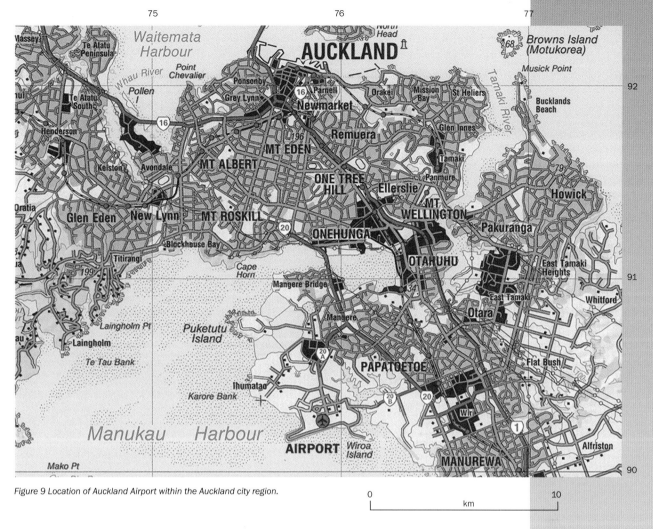

Figure 9 Location of Auckland Airport within the Auckland city region.

Auckland airport is located about 20 km south of the city centre on flat land next to the Manukau Harbour. The location is towards the edge of the city but urban expansion and infill of farmland means the airport is today very much within the city boundary rather than on the urban fringe as it once was (Figure 9).

Airport location reflects a balance between opposite forces (Figure 10)

- **Within the city benefits** The closer an airport is to the city centre, the more benefits result because of shorter travelling times from the airport to the central area of activity and the rest of the city. A time of 60 minutes is often seen as the maximum time it should take to reach the airport from all parts of the city. Beyond the threshold of an hour, an airport does not serve its city area well because of the large amount of time that must be spent to reach it. If an airport is part of the city rail system, then the problem of distance is reduced. The main airports of Paris and Hong Kong are examples of airports where rail connections overcome the problem of distance away from the city. However, other airports such as Narita/Tokyo and JFK/New York have poor connections with their city areas because of congestion and the lack of alternatives to road access.

- **Within the city costs** The closer an airport is to the urban area and city centre, the more costs and problems increase. Land within the city is usually more expensive than land on and beyond the edge of the city. Airports and infrastructure that goes with them require a large amount of land. Land costs for building an airport within the city are therefore very high. Airports taking up lots of city land also means the chance for using that land for other purposes like housing, parks, shopping and commercial centres is lost. The closer the airport is to residential areas, the greater the number of people there will be affected by airport noise. Airlines themselves may face restrictions on after-dark operations because of noise they create.

The most suitable zone for airport location (pink-shaded area) is a compromise between being near to the city centre and being far away. Auckland Airport can be viewed as being within the zone of 'most suitability'.

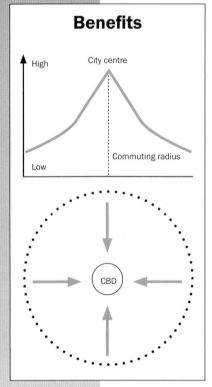 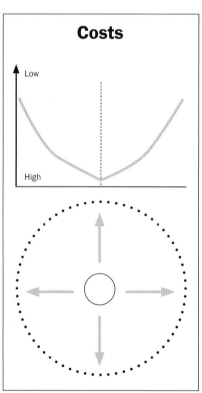 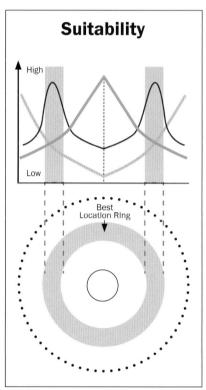

Figure 10 Factors that influence airport location — a balance between two opposite forces.

Auckland Airport — New Zealand's major domestic and international airport

14 million travellers each year:
a relatively even split between international and domestic

24 million
passengers a year – the forecast for 2025

36,800
travellers every day

868,000
transit passengers a year

154,971
aircraft movements per year

Around **70%**
of all international visitors to New Zealand arrive or depart from Auckland

200,000 plus
tonnes of high-value freight

Largest airport in New Zealand
Second largest in Australasia – just behind Sydney

22 international airlines
connect to 33 destinations

Figure 11 Auckland Airport in statistics — a 2013 snapshot.

Rank	Airport	Passenger numbers
1	Auckland	14,200,000
2	Christchurch	5,500,000
3	Wellington	5,200,000
4	Queenstown	1,160,000
5	Dunedin	854,000
6	Nelson	777,000
7	Palmerston North	450,000
8	Hawke's Bay	441,000
9	Hamilton	354,000
10	New Plymouth	322,000

Figure 12 New Zealand's top 10 airports for passenger numbers 2012 (international and/or domestic).

The numbers in Figures 11 and 12 for passenger and freight movements are large in themselves. These people and freight numbers connected with the airport become even larger when additional numbers are included:

- In terms of the value of cargo handled, Auckland Airport is the second largest port in New Zealand. Only the seaport of Auckland imports and exports a higher value of goods.
- There are, on average, two people coming to the airport for every person flying — these are the people dropping off and picking up visitors, friends and family who are flying in and out of the airport.
- Thousands of people work in and around the airport in jobs associated directly (e.g. baggage handlers and flight controllers) and indirectly (e.g. in freight movement companies and car rentals) with the airport and nearby business park area. Most of these people commute daily to the airport area for work.
- The airport and adjacent airport-linked businesses, shops and services are like a mini-city and on their way to being an aerotropolis (Figure 4).

Figure 13 Terminals and linked commercial area.

Figure 14 Airport take-off.

Figure 15

	Present number 2013	2044 projection
Passenger total per year (domestic/international)	14.5 million (6.7/7.8)	40 million (16/24)
Passengers using airport each day	40,000	110,000
Aircraft take offs and landings per year	155,000	240,000
Employment at airport and in surrounding area	20,000	60,000
Daily vehicle trips to and from airport	63,000	140,000
Public car park spaces	6100	15,900
Staff car park spaces	2500	6300

Figure 16 Future growth projections.

Learning Activities

1   Refer to Figure 8. Design a poster titled 'Airport of the future'.

2   Refer to Figures 9 and 10 and the text.

   a   Describe the location of Auckland Airport.

   b   What makes this a good site for an airport?

3   Refer to Figures 11 and 12. Choose and list three pieces of evidence to support the statement that 'Auckland Airport is New Zealand's major domestic and international airport'.

4   Refer to Figures 13–16. Why is access to Auckland Airport going to be an ongoing issue between now and 2044?

Figure 17 Auckland Airport — aerial photo.

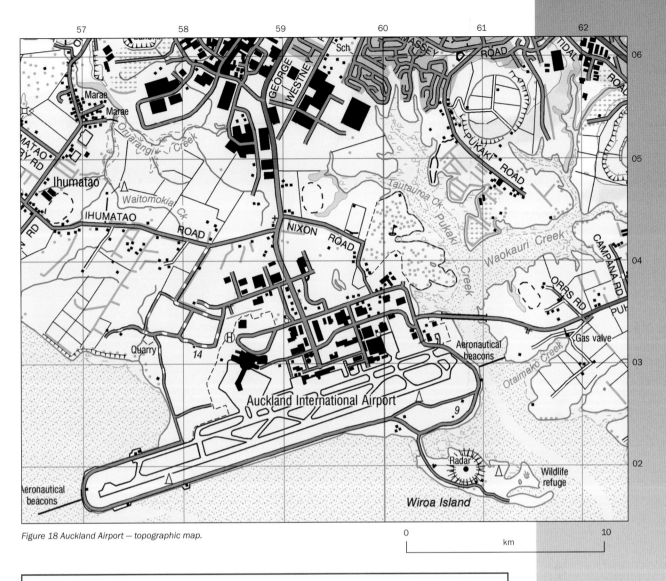

Figure 18 Auckland Airport — topographic map.

ISBN: 9780170233316

Learning Activity

1 Refer to Figures 17 and 18. Draw a precis map to show the coast outline and airport location. Mark on the map five other features of the airport site and surrounding area — two of these could be mangroves and business area.

The airport — connections and accessibility

Apart from air travel itself, the main access to Auckland Airport and the adjacent airport commercial and business area is by road. The airport is surrounded by public and staff car parks (Figure 19), while within close distance of the airport are new large 'Park & Ride' car parks where cheaper long-term parking is available. Free shuttle bus and van services connect these car parks with the airport terminals (Figure 20). A free shuttle bus also runs between the international and domestic air terminals.

Most people access the airport by road using one of three types of transport:

- Private car and rentals (Figures 19 and 23)
- Taxis and shuttle vans
- Public buses. An Airbus Express service runs a regular service 24 hours a day (Figure 21). This service connects the airport with the central city transport hub at Britomart where bus and train connections to the wider Auckland area can be made as well as giving access to the Downtown Ferry Terminal just across the road from Britomart. Other public buses provide connections between the airport and transport centres in South Auckland, where connections to the rail and Auckland bus network can be made (Figure 22).

Figure 19 Public car park.

Figure 20 'Park & Ride' bus.

 ISBN: 9780170233316

Figure 21 a and b Airbus Express.

Figure 22 Local bus connections.

Figure 23 Private cars and rentals.

Most people travelling to and from the airport by road use a combination of Auckland city streets and motorways. Currently, there is no direct motorway access to the airport, and at some point, airport traffic must use Auckland city streets. Two state highways provide connections to motorways that give road access to and from the airport. State Highway 20A provides links to the north and to the west: to central Auckland, the western suburbs and to the North Shore and then on to Northland. State Highway 20B leaves the airport to the east and provides access to southern and eastern Auckland, and then the rest of the North Island (Figures 9 and 24).

Travel from the city centre to the airport by road takes around 45 minutes when roads are clear. Congestion at the beginning and end of the business day and during peak times for flight arrivals and departures can mean gridlock and travel times of 90 minutes. When road crashes take place, traffic chaos occurs on the access roads.

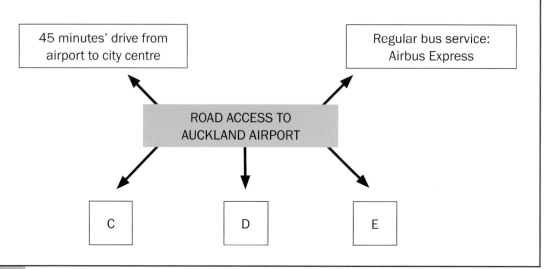

Learning Activity

1 Refer to the section entitled 'The airport — connections and accessibility' on pages 217–219. Complete this star diagram titled 'Road access to Auckland Airport'.

| 45 minutes' drive from airport to city centre | | Regular bus service: Airbus Express |

ROAD ACCESS TO AUCKLAND AIRPORT

| C | D | E |

Where would an airport rail link be built?

This is a matter of location and route choice. Three options are available (Figures 24–26).

1 a link from the existing rail branch line and terminus at Onehunga
2 a link to the main line in Papatoetoe
3 a circular route linking to both Papatoetoe and Onehunga.

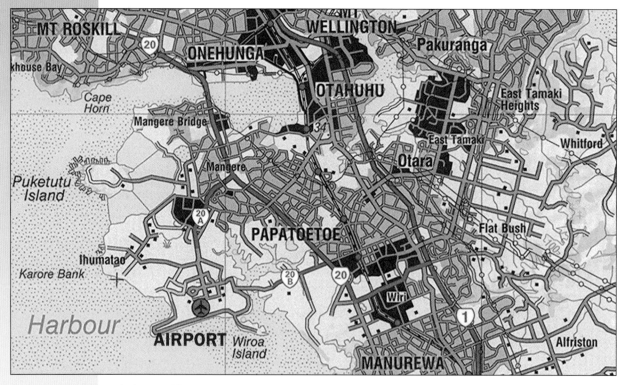

Figure 24 Auckland Airport, Onehunga and Papatoetoe — local area transport links.

0   km   10

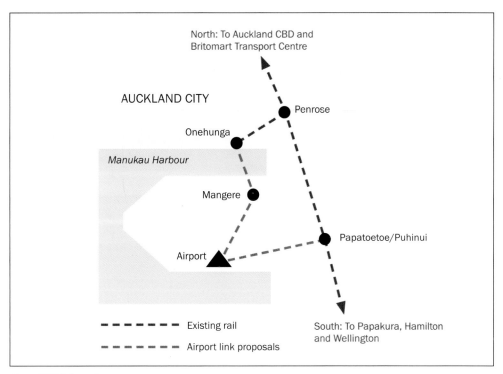

Figure 25 Rail line options.

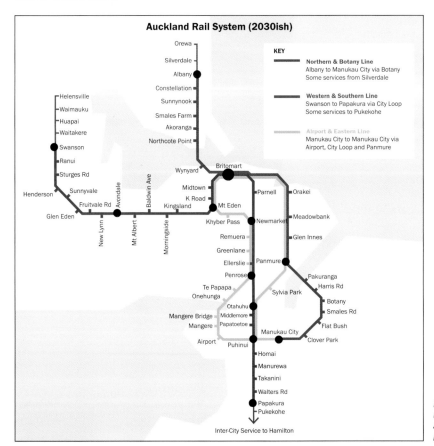

Figure 26 An airport rail-link, part of future Auckland rail network proposals.

Learning Activity

1 Make a copy of the map in Figure 25. On your map, add labels to identify the three possible airport rail-link options. Put distances on each section of the route (Onehunga – Mangere; Mangere – Airport; Papatoetoe – Airport). Use Figure 24 to help with this task.

Different views about the issue

A Auckland City Mayor Len Brown has promoted and been a supporter of an airport rail link for a long time. He did so when he was mayor of Manukau City (this is where the airport is located) and continued to do so when he became mayor of the new Auckland supercity. He says studies show that to provide efficient transport to and from the airport, a rail link needs to be part of the mix. He has argued that not only is the airport the gateway into New Zealand for tourists and business travellers from overseas, but the airport business and commercial corridor is expanding quickly and would also benefit from a rail link.

B 'Developing Auckland Airport — city centre rail link is a no-brainer'. This was the message given by a keynote speaker at an Auckland business conference. Delegates at the conference were given an illustrated presentation where they were told that modern cities and modern airports have city–airport rail links as part of their design. Sydney, for example, has an airport–city centre 'Sydney Train' that runs every 10–15 minutes. For a fare of less than $20 you can be at the airport or be in the city centre in only 30 minutes. The new Hong Kong airport has a similar efficient, modern rail link to the city centre (journey time of 25 minutes for a 35 km distance, costing about NZ$18 one way). Using rail makes for a much better flow of people to and from the airport than trying to use congested roads. The problem is that travellers want to get to the city centre, which is already the most congested part of the city. Airport buses and shuttles and people using private cars to travel to and from the airport only adds to this congestion. Putting more people onto roads, whether in cars or buses, as airport traffic increases makes no sense. Having a city–airport rail link is the way to future proof the airport.

Figure 27 Modern trains and stations providing a connection to a 21st-century airport is one vision of the future.

C Auckland Chamber of Commerce: From the perspective of our business community, transport is the number one issue facing Auckland. Many surveys also show the general public of Auckland share this view. A fast and efficient rail network for Auckland makes sense but it will be expensive to build and run. Early completion of the inner city CBD rail loop is our number one rail priority — this will allow for a much more efficient use of the existing rail system. This should mean more people will use rail to travel and road congestion problems should ease because of this. Improved ferry services, continuing road improvements and focusing on safety for cyclists by building a network of cycleways are also priorities that will compete for finance along with a rail link to the airport. The airport rail link may seem like a good idea now but may not seem so good when all the building costs are presented. Central government finance needs to be guaranteed for the airport rail link and a strong case can be made for such finance as the airport and its business park are national assets, not just Auckland ones.

Figure 28

D A 'Promoting Auckland' online group have come out in favour of an airport rail link, but are concerned about impact on rates. The group have said, 'We strongly favour an airport rail link and think a service with new stations with modern electric trains would be attractive for passengers and mark Auckland as a dynamic 21st-century city. It would help attract business people and tourists to the city where we are competing with places like Sydney and Melbourne for such visitors. However, we don't want further rates increases to pay for rail. The new trains are already bought and coming into service, so private sector investment is needed (maybe a large overseas company) to pay for the cost of new station construction plus ongoing running costs and share the line-building costs with local and central government. In return, the company could take profits from the running of the airport line over a twenty-year period.'

E Taxi drivers and shuttle bus drivers have expressed alarm at any rail link to the airport being developed. They argue that they have trouble already making a decent wage and that airport custom often makes up over half their weekly income. They say that a rail link between the CBD and the airport would put them and their companies out of business. Hence they are very much against the proposal.

F Consultants who have reviewed future growth of passenger numbers and freight at Auckland Airport have not favoured building a rail link. They argue that:
  i There would be a lot of local community opposition to any rail-link build whether it be one way between either Onehunga or Papatoetoe or a circular route. All these links would require a double line and take up a lot of land. Much of this land is already residential land and community disruption and destruction would result in strong opposition. Locals would feel their interests were being sacrificed in favour of those of the outsiders — the tourists and other travelling public.
  ii Roads are coping now and improvements are planned. Cycle paths are also part of future planning proposals and although these would cater for only a few tourists, they would be an attractive alternative to using cars and public road transport for airport workers.
  iii The future growth of passenger and freight traffic through Auckland Airport seems certain. The growth figures, however, ignore the fact that many passengers arriving and departing on international and domestic flights never leave the airport — they are in transit and simply transfer from domestic to international and vice versa. Freight growth cannot justify a rail-link build either. Much of the freight is light but high value and is better suited to being transported by road (vans and small trucks) rather than by rail.

G Rapid rail has been identified as the best long-term public transport option in an investigation by Auckland Transport for improving transport in Auckland's airport/southwestern area. The investigation showed that in addition to growing passenger numbers using the airport, this area of southwest Auckland around the airport has become a major employment centre with transport, commercial, retail, recreational and residential areas. Transport improvements have been identified as being needed to serve the needs of these local communities as well as for travellers to and from the airport. The airport rail link itself was listed at number four in projects to provide good access and transport options in the airport/southwest area. The projects were listed as needing completion over a 30-year time span.

- In the short term complete vital roading improvements.
- Make improvements to public transport such as local buses.
- Provide better cycling provisions.
- In the longer term, build an airport rail link to connect with the rest of Auckland and North Island network.

H Our business is based on the edge of the Auckland CBD in Ponsonby. We also have staff in Napier and Wellington. When they fly up to Auckland, as they do about once a month, they have to get a taxi from the airport because it's just too time-consuming to get the bus. The taxi cost is $75–90 each trip (one way). A rail link could get them to the CBD (Britomart transport centre) quickly and cheaply and then it's a short taxi ride from there to the office. We will be far from alone in that situation. It's probably tens of millions of dollars a year being wasted on taxis because there's no reliable, quick public transport option for people who often have luggage (business people as well as tourists) and need to get from the airport into the CBD and surrounds. New Zealand is supposedly a leading developed country, yet we have no rail link to our biggest airport. We are the laughing stock of the rest of the world when people get off their plane and can only get on a noisy, smelly bus, into a cramped shuttle van or expensive taxi. Rail is the answer — come on, people, wake up! A line to the airport should be the top transport priority for Auckland.

I The 'Campaign for Better Transport' are frustrated that although there remains strong public support for the building of an airport rail link, with petitions having been presented to council and government, no real progress has been made. Words of support from previous councils and governments have not been followed by any action. The project seems to have dropped down the list of council priorities, a Campaign spokesperson said. In their meetings with the new council, other transport projects like roading, cycle paths and the costly CBD rail loop project have been put ahead of any airport rail link. The Campaign group say that future-proofing transport in the airport area with the building of a rail link needs to happen now. If we wait until all the new passengers start arriving and more businesses open in the airport corridor area, it will be too late, they say. The rail project is a long-term one that needs starting now, the group say, and they conclude: 'Passenger numbers through Auckland Airport are forecast to triple to 40 million by 2041. There are already around 900 businesses in the airport business precinct, and the adjacent area is an emerging major employment centre with transport, commercial, retail and recreational as well as residential use. The beauty about expanding rail to the airport is that it won't just provide for people flying in and out of the airport with a reliable connection to the city, but it will also provide thousands of commuters in the airport region with an alternative to their cars. Traffic congestion is only going to get worse so we need to be expanding our rail network and its capacity as a matter of urgency.'

Learning Activities

1 **a** There are nine views on pages 222–224 (A–I) given from groups and individuals about whether or not a rail link to the airport should be built. Some are for the building of a link, some are against and others are more in the 'not certain' category.

Make a copy of this opinion tower and then put letters A–I into each square based on studying the information presented about each of the nine views.

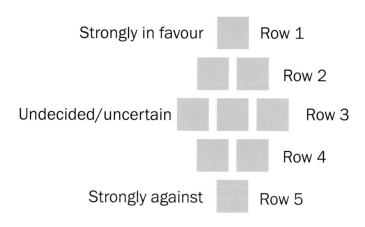

Strongly in favour — Row 1

Row 2

Undecided/uncertain — Row 3

Row 4

Strongly against — Row 5

b Choose five letters from the tower — one from each row. For each of these letters, explain why the group or person holds that view about the rail-link building.

Views change over time

- Some transport planners who have been very supportive of an airport rail link in the past have changed their mind. They say that as motorways and other suburban road projects are completed, the capacity to move people around Auckland (including to and from the airport) will increase. They point to evidence from the North Shore area of Auckland where the introduction of frequent and fast bus services using special bus lanes led to a huge increase in the numbers of commuters travelling by bus into the CBD each day. Cars were left at home or in 'park and ride' car parks. Traffic congestion was reduced. These transport planners say that having a network of similar bus-only traffic lanes between the CBD and the airport would make a bus option to the airport more attractive. People would be encouraged onto airport buses not just because of the convenience of travel but also because airport parking costs would be saved. With easy airport access by road, the need for a rail link with all the costs involved would fade away.

- Councils and central government who have previously put an airport rail link on a list of priority future projects now place the link on a 'wish list'. High cost of the rail build is one reason but it is all about costs and benefits in the bigger sense — other transport projects are seen to make more sense. Completion of the Auckland motorway network, building the CBD rail loop and a second harbour bridge (or tunnel) are all seen as having better cost-benefit outcomes than an airport rail link. In times when money is tight, the rail link has slipped down the list of 'must do' projects. The economic argument for other transport projects has been seen to be greater.

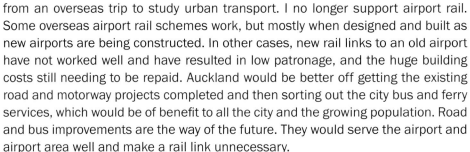

Figure 29 Airport links — rail or road?

- **Opinion:** I am an Auckland ratepayer and member of 'Promoting Auckland'. I have a freight forwarding business near the airport (in the new airport business zone) and have just returned from an overseas trip to study urban transport. I no longer support airport rail. Some overseas airport rail schemes work, but mostly when designed and built as new airports are being constructed. In other cases, new rail links to an old airport have not worked well and have resulted in low patronage, and the huge building costs still needing to be repaid. Auckland would be better off getting the existing road and motorway projects completed and then sorting out the city bus and ferry services, which would be of benefit to all the city and the growing population. Road and bus improvements are the way of the future. They would serve the airport and airport area well and make a rail link unnecessary.

- Families and Community First: Local community groups in the Mangere and Papatoetoe area have long supported the airport rail-link project. It would improve their access to the wider Auckland region. Many of the people work in the airport precinct and the trains would give them a better way to get to work than travelling by car or using local bus services. The communities also see they would benefit from being able to use the train going away from the airport — if they work in the city they could take the 'airport train' to get to work, school and university and also the service would give them better access to shopping malls like Sylvia Park, which has a rail link to it. The general community view in favour of the rail project changes, though, when detailed route proposals are put forward. People living in areas along the route whose homes would be demolished by a new rail line or would have their community split in half by any line become opposed to the proposal. It becomes the neverending problem of 'support for big projects in principle' becoming opposed when their lives and community face disruption.

Learning Activity

1 Each of the four groups/people above have changed their view of support for an Auckland rail link to one that is opposed or at least less positive. Why have these changes of view come about?

Figure 30

Should an airport rail link be built? Weighing up the evidence and making a decision

The case for
- There is a case for a rail link today based on the large number of people travelling to and from the airport area by road. Not only would it give these people transport options but it would also ease traffic congestion on airport access roads.
- Any rail link between the airport and existing lines would give a link to the city rail network and a direct link to the CBD Britomart transport centre for onward rail, bus and ferry connections.
- The case for the rail link becomes even stronger when future growth at and around the airport (Figures 11 and 16) is taken into consideration. Much of this growth can be seen happening now as building and development projects continue in the airport precinct.
- It is not just about travellers. A rail link would be able to move freight and goods to and from the airport and airport business complex and also provide another public transport option for the large number of people who work in the airport area.
- Overseas airports with efficient rail links to their city centres like Sydney, London and San Francisco are often quoted as being models that Auckland should follow.

The case against
Costs would be huge. Estimates are as follows:

- Option 1 The link from Onehunga — $1000-2000 million.
- Option 2 The link to the main line in Papatoetoe — $500 million.
- Option 3 The circular route linking to both Papatoetoe and Onehunga — upwards of $2000 million.
- Practicality. Any rail link will require a large amount of land and much of the land is already built over with residential housing and commercial premises. Lives and communities would be disrupted (Figure 30).
- There may be engineering issues with a harbour crossing needed from Onehunga and estuary land to cross with any link to the main north–south rail line in the Papatoetoe area (Figures 24 and 25).
- Necessity. Existing roads can and do cope with the airport traffic, and continuing improvements to these roads and to the bus shuttles make a rail link unnecessary.

1 **a** Based on the information provided so far in this chapter, draw a seesaw diagram showing the case for and the case against the building of an airport rail link. Tip the seesaw to show your view on whether the link should be built. For example, if you judge that the case for outweighs the case against, then tip the case against end downwards.

b Justify the way you have tipped the seesaw.

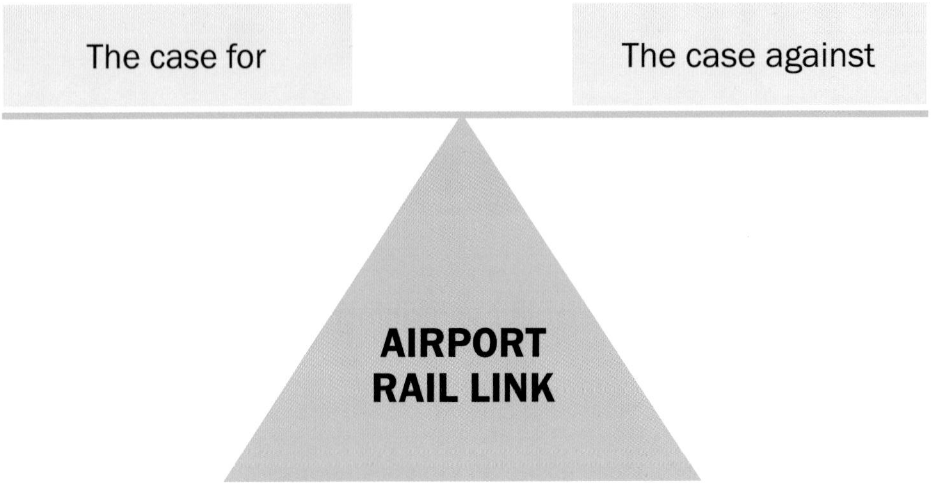

The case for

The case against

AIRPORT RAIL LINK

2 Put the information from Figure 6, 'Airport geography', into a table. For each of the five reasons listed, give examples or provide elaboration with information that is related to Auckland Airport. Here is a start:

Geographers interested in	Auckland Airport example/elaboration
Location	It is located on a large area of flat land next to the Manukau Harbour. The airport location is about 20 km south of Auckland city centre. The journey to the city by road from the airport takes about 45 minutes.
Movements and connections between places	
Movements and connections between people	
Accessibility	
Growth and change	

3 Prepare a visual display for a public meeting (poster or PowerPoint) that would give information about the following. Remember that you are taking the role of the expert. You need to inform the meeting in a clear, key-point coverage (not too much detail) and interesting way. Use the photos like the ones on the following page to help illustrate your presentation.

a The location and setting of the airport rail-link issue.

b What the issue is about and the options there are.

c Possible courses of action and the consequences of these options.

d How the best course of action might be determined.

e Next steps that could take place and be followed.

Learning Activities

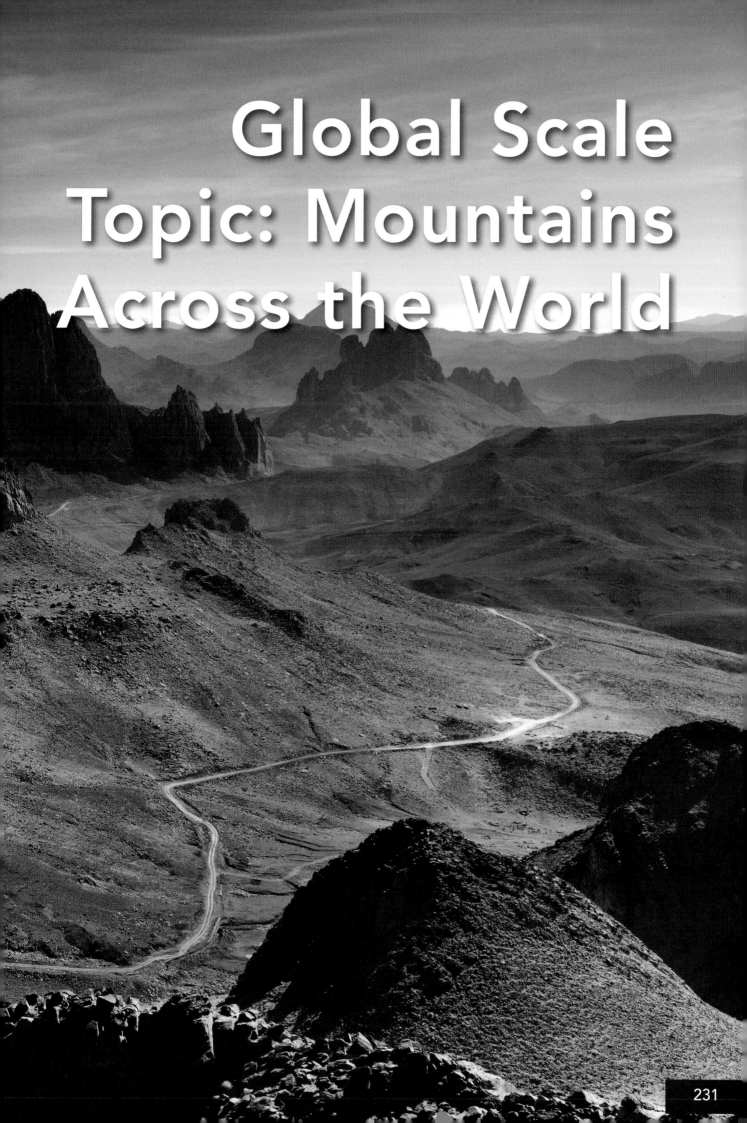

Global Scale Topic: Mountains Across the World

6 World Mountain Areas

Mountain areas across the globe – an overview

There is no one definition of what makes a mountain or a mountain range. A common definition of a mountain is an area over 1000 metres in height above sea level with steep slopes. Mountain ranges are areas of continuous elevated (high) land that contain many individual mountain peaks and cover large areas of the earth's surface. Using this definition, mountains and mountain ranges cover about 25 percent of the surface of the earth.

The world's longest mountain range is the Andes on the western side of South America. It is over 7000 km in length. The Himalaya-Karakoram range in Asia is the world's highest mountain range. The highest 30 mountains in the world (all over 7500 metres in height) are all located within this range, including the world's two highest and most famous mountains, Mount Everest (8848 metres) and K2 (8611 metres).

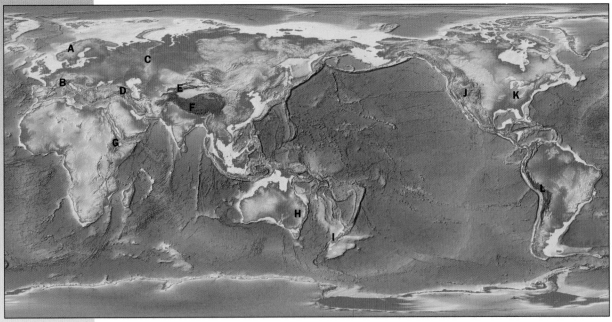

Figure 1 Mountain areas of the world.

 ISBN: 9780170233316

Figure 2 Mount Kilimanjaro, Africa.

Figure 3 Mount Everest, Asia.

The highest mountain in each continent

Continent	Highest mountain	Height (metres)	Country location	Range within which the highest mountain is located
Asia	Mount Everest	8848	Nepal/China	Himalayas
South America	Aconcagua	6962	Argentina	Andes
North America	Mount McKinley	6194	USA	Alaska
Africa	Mount Kilimanjaro	5892	Tanzania	Kilimanjaro
Europe (including Russia)	Mount Elbrus	5642	Russia	Caucasus
Western Europe excluding Russia	Mont Blanc	4810	France/Italy	Graian Alps
Antarctica	Vinson Massif	4892	n/a	Ellsworth
Australasia	Puncak Jaya	4884	Indonesia (Papua province)	Sudirman
For comparison – the highest mountains in Australia and New Zealand				
Australasia	Aoraki/Mount Cook	3754	New Zealand	Southern Alps
Australasia	Mount Kosciuszko	2228	Australia	Snowy Mountains

Figure 4

Learning Activities

1 Write a definition of a mountain area.

2 What size or height record do each of the Andes and the Himalayas hold?

3 The map in Figure 1 shows the location of mountain areas across the world. Copy the table of mountain area names and, using an atlas to help you, correctly match the letters A–L from the map with the mountain area names in the table. The first one has been done as an example.

European Alps **B**	Caucasus Mountains	Ural Mountains	Great Dividing Range
NZ Southern Alps	Ethiopian Highlands	Tian Shan Mountains	Scandinavian Mountains
Appalachians	Himalaya-Karakoram	Rocky Mountains	Andes Mountains

4 Study Figures 2 and 3.

   a Compare and contrast: In what ways are Mount Kilimanjaro and Mount Everest, including the areas around each mountain, **i** similar, **ii** different?

   b Suggest reasons for the differences — a check on the global location of each mountain would be a good way to start this answer. Also consider the shape of each mountain and the processes that would have led to the formation of mountains of these shapes.

5 Construct a graph (e.g. a column/bar graph) to show the heights of the mountains in Figure 4. Follow all graph presentation conventions.

Location and global pattern

Key terms

Global refers to the whole world. A global study would be one that included places and countries from different continents and from different hemispheres.

Location describes the position of something. For example, three ways of describing the location of a mountain would be naming the country the mountain was found in, or giving the latitude and longitude of the mountain, or stating how far the mountain was from a well-known place.

Pattern is about the way things are arranged and the shapes of things. For example, studying the location of mountains across the world shows that mountains are often found close to each other in the same area. This would be called a clustered (grouped together) pattern.

Spatial means related to the 'space' of the surface of the earth. When the location and position of things on the surface of the earth are studied, patterns can often be seen. The clustering of mountains is an example of a **spatial pattern**.

Important understanding: Mountains and mountain ranges have an uneven distribution pattern across the globe, but the location of the mountains and mountain ranges is not random.

If the world is divided into natural regions like mountains, forests, ice-covered areas, grasslands and deserts, the mountain regions are distinctive because they occur in every continent, at all latitudes, and in all of the world's main environment areas (Figure 1). Mountain regions can appear to be unevenly and randomly spread across the globe. They are not located in particular latitude bands like other natural regions such as deserts and rainforests.

Alignments and locations of mountain ranges vary from continent to continent.
- In Oceania, the New Zealand Southern Alps have a northeast–southwest alignment.
- In North and South America, mountain chains that include the Rockies and Andes run north to south along the western side of the continents. They begin in the cold regions of Alaska above 60°N, run down through the tropics of Central America, on across the equator in Ecuador, and end as far as 50°S in the southern parts of Chile and Argentina.
- In Asia, the great ranges of the Himalayas are well inland and extend west to east.
- The European Alps are also inland and have a south-to-north orientation, where they form the border between France and Italy and then extend west to east across Switzerland and into Austria.

Although mountains and mountain ranges have an uneven distribution pattern across the globe, within this overall global pattern other patterns like groups of high mountains (clusters) in some areas and lines of mountain ranges (linear pattern) in other areas can be identified. When the formation of mountains and mountain ranges are studied, another point about their locations becomes clear: the locations are not random but are linked with the location of the processes that led to their formation.

There are two great global belts of mountains (a belt is a large area that is usually wide and long in shape) within which most of the world's great mountain ranges and mountain peaks are located. These belts form curved linear (line) patterns across large parts of the globe (Figure 5).

1 A circular belt of mountains located around the outer edge of the Pacific Ocean (sometimes called the Pacific Ring of Fire). This belt includes the mountains of New Zealand, the western coast of the Americas (Andes and Rockies), Kamchatka Peninsula in Russia, Japan, Taiwan, the Philippines, as well as parts of Indonesia and Papua New Guinea.

2 The Alpide Belt is a huge belt of mountains located on the southern edge of Eurasia (Europe and Asia). The mountains stretch from Indonesia to the Himalayas and the other mountains of central Asia, and then westwards through the Middle East to the European Alps and on to the mountains of southern Spain and Atlas Mountains in North Africa.

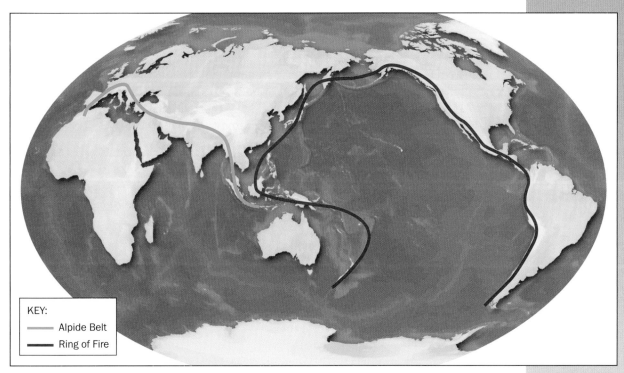

KEY:
— Alpide Belt
— Ring of Fire

Figure 5 Mountains of the Pacific Ring of Fire and Alpide Belt.

There are other large mountain ranges outside of these two great belts.

Africa

In addition to the Atlas Mountains in the north, there are two other major mountain areas.

1 The mountains that run north to south on the eastern side of the continent, which include the Ethiopian Highlands and other mountains associated with the Great Rift Valley ending with the Drakensberg Range in South Africa (Figure 6). These mountains are caused by plate movements and faulting. Some of the highest mountains, e.g. Mount Kilimanjaro (Figure 2), are volcanic in origin.

2 The massive isolated mountain blocks in the Sahara Desert (the Hoggar and Tibesti Mountains, Figure 6). These are examples of volcanic hot-spot mountains (see Case Study 1 on page 243). The highest peaks in these Sahara mountains are between 2500 and 3500 metres in altitude.

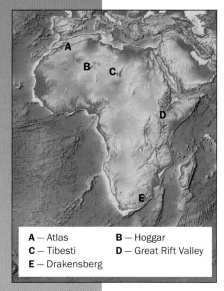

A — Atlas   B — Hoggar
C — Tibesti   D — Great Rift Valley
E — Drakensberg

Figure 6 Mountain ranges in Africa.

Figure 7 Hoggar Mountains in North Africa.

Europe, Australia, the USA and Russia

In northern Europe, the Scandinavian Mountains run the length of the west side of the Scandinavian peninsula in Norway and Sweden. The Great Dividing Range in Australia also runs lengthwise along the eastern side of that country, as do the Appalachian Mountains in the east of the USA. In Russia, the Ural Mountains have a north–south orientation following close to longitude line 60°E.

Learning Activities

1 **a** Refer to Figure 5 and on an outline map of the world, mark and name

  i   the Alpide Belt,

  ii   the Pacific Ring of Fire.

  b Using an atlas and Figure 1 to help you, label within the belts the location of these mountain ranges:

  European Alps

  Caucasus

  Himalayas

  Rockies

  Andes

  New Zealand Southern Alps.

2 When referring to mountains, what does the word 'belt' mean?

3 What pattern best describes the shape of the two great mountain belts of the world?

4 Write a paragraph titled 'The Alpide mountain belt'.

5 Identify each of these mountain ranges.

  a A mountain range in North Africa that is part of the Alpide Belt.

  b The mountain range that forms the border between France and Italy.

  c The mountain belt that the New Zealand Southern Alps are located within.

  d They run north to south on the western side of South America.

  e Two mountain ranges located in the Sahara Desert.

  f Located close to longitude 60°E in Russia.

  g Run parallel to the east coast of Australia.

  h Located in the east of the United States.

Learning Activities

6 Copy the paragraph and insert the correct words where the red dots are placed. Choose from the word list shown.

Mountain areas are unevenly ● *across the world. Mountains are found in all continents and in all* ● *areas. Some mountain areas are inland, others are* ● *to the coast. The Himalayas, the world's highest mountain range, are* ● *in the interior of Asia. The Andes Mountains, on the other hand, are located close to the* ● *coast of South America. Most of the world's great mountain ranges and mountain peaks are located within either the* ● *area or the Pacific Ring of Fire, which are the two great mountain* ● *of the world. The Hoggar and Tibesti Mountains in the Sahara Desert are examples of mountain ranges that are outside these two belts. Mountain ranges that are part of the Pacific* ● *are located around the* ● *of the Pacific Ocean. These mountain ranges form a semi-circular* ● *and peripheral pattern. This semi-circular belt includes mountains from four* ● *: Oceania, Asia, North and South America. When viewed as a whole, the different mountain ranges form an* ● *but distinctive global pattern.*

Word list

distributed	Alpide	linear	located
uneven	edge	Ring of Fire	continents
close	environment	belts	west

Background — the crust of the earth, plate tectonics and mountains

Important understanding: Mountain ranges are located in belts along or near plate boundaries.

To understand how mountain ranges are built, you need some background on the earth's crust and plate tectonics. The word '**tectonics**' has Greek and Latin origins and means 'building' or 'construction'.

The outside solid shell of the earth is called the **lithosphere**. The thin and upper part of the lithosphere is the **earth's crust**. The lower part of the lithosphere is formed of rocks of the top part of the **mantle** that are brittle and rigid. The lithosphere does not form a single continuous shell over the earth but instead is broken into pieces called **plates**.

There are two types of rocks that make up the crust: crust that is beneath the oceans that is mostly made of **basaltic rock**, and crust that forms the continents that is mainly made of **granitic rock**. The **ocean crust** forms a layer over the whole of the outside of the earth. **Continental crust**, which makes up the land areas, is found only in some locations. Where it does occur, it rests on top of the ocean crust because it is less dense (lighter) than the ocean crust (Figures 8 and 9).

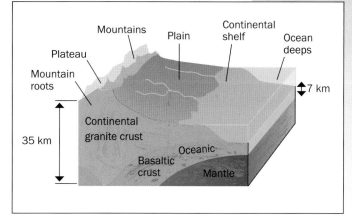

Figure 8 The crust of the earth.

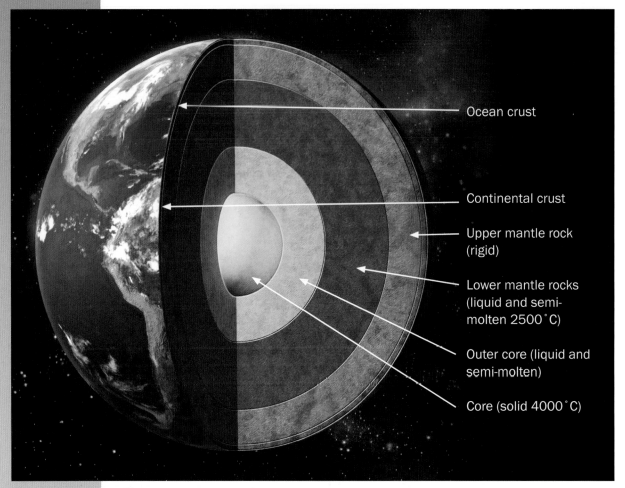

Ocean crust

Continental crust

Upper mantle rock (rigid)

Lower mantle rocks (liquid and semi-molten 2500°C)

Outer core (liquid and semi-molten)

Core (solid 4000°C)

Figure 9 The interior of the earth.

The plates of the earth's outer shell are in **continual movement** powered by energy sources from the core and lower mantle area of the earth's interior. The movements are at speeds that are very slow in human time — just a few millimetres to several centimetres each year, about the same speed that a fingernail grows. Measured in geology time of thousands and millions of years, these small yearly movements can cause movements over big distances — a movement of 30 mm each year would mean a movement of 300 metres in 10,000 years, 3 kilometres in 100,000 years and 30 kilometres in a million years.

At some plate boundaries, **new earth crust** is being created as material from the mantle area breaks through onto the surface. This happens at plate boundaries beneath the sea at **mid-ocean ridges**. The ridge along the middle of the Atlantic Ocean is an example of such a boundary. These types of plate boundary are called **divergent** because plates are moving apart from each other. If this adding of new material happened at all plate boundaries, the earth would be growing bigger in size. However, the earth is not growing bigger because at other plate boundaries crust is being destroyed. This is where plates are moving together and colliding. These boundaries are called **convergent** boundaries, and it is here that mountain building takes place.

Because plate movements are powered by forces from within the earth, they operate in ways that are independent of surface conditions. It is for this reason that mountain ranges are not located in regular bands across the surface of the earth. Mountains are as likely to run north to south as they are east to west. Volcanoes are as likely to occur in cold Arctic areas as they are in hot equatorial regions. Mountain ranges are located in belts of crustal collision along convergent plate boundaries.

Relationship: mountains and convergent plate boundaries

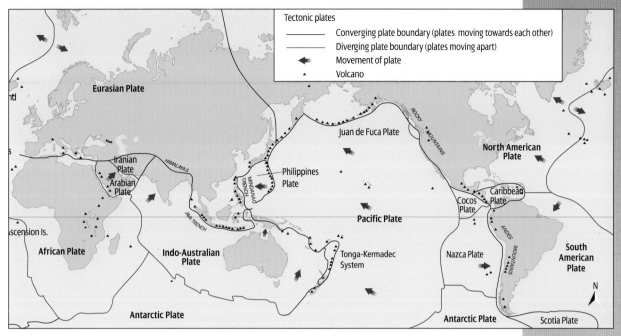

Figure 10 Crustal plates and crustal boundaries. Mountains form at convergent boundaries.

Learning Activities

1   Write sentences that give the meaning and explanation of each of these words:
    • tectonics
    • crust
    • plate
    • convergent
    • mantle.

2   What causes the crustal plates to move?

3   Crustal plate movement is 'very slow but has huge consequences'.
    a   How slow is 'very slow'?
    b   What are the 'huge consequences' of this very slow plate movement?

4   a   Make a copy of Figure 8.
    b   What proof is there that continental crust is lighter than ocean crust?

5   Copy the sentence that best describes the location of the mountain areas of the world.

6   a   Refer to Figure 10 and on an outline map of the world, mark the location of:
        i   convergent plate boundaries
        ii   mountains
        iii volcanoes.
    b   Name five of these mountain ranges/volcanic areas.

Causes of the global pattern – why are the mountains located where they are?

Important understanding: Mountains have been formed more in some parts of the world than in other parts. In places where tectonic processes and volcanic activity take place, mountain formation has occurred.

Plate tectonics and volcanic activity lead to the formation of mountains. Both of these are the result of forces taking place in the interior of the earth that then affect the surface features (Figure 11). These internal forces result in the formation of mountains where plates push against each other causing parts of the earth's crust to be compressed and then folded, faulted and forced upwards. Volcanic eruptions also occur in these areas. Most of the world's mountain ranges are located on or close to plate boundaries. Plate boundaries are places where tectonic processes and volcanic activity result in mountain formation. Therefore the location of plate boundaries has a big influence on the location and distribution pattern of mountain areas.

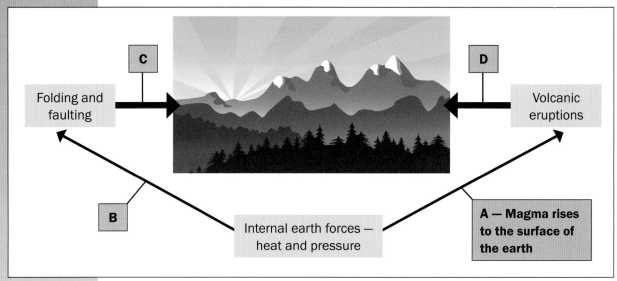

Figure 11 Internal forces and mountain formation.

Mountain formation

Where two plates move towards each other, a slow-motion collision between the plates takes place. Colliding plate boundaries are known as converging (coming together) boundaries. Mountain formation occurs in these zones of convergence. There are different types of converging boundaries.

A **When two plates of ocean crust converge**, one plate gets forced below the edge of the other plate. This forcing down process is called subduction. As the subducted crust is dragged deep into the earth, some of it melts. The melted (molten) rock is called magma. Pressure within the earth's interior causes some magma to move upwards through weak areas of the crust. Volcanic eruptions then occur on the surface of the earth and chains of volcanic mountains are formed. The islands and mountains of Japan and the Philippines were formed in this way.

 ISBN: 9780170233316

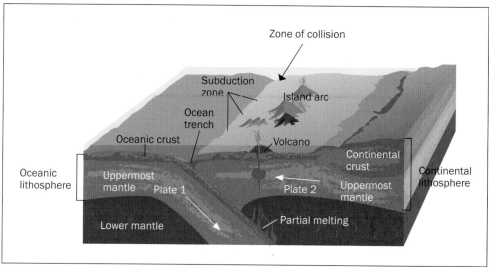

Figure 12 Mountains form where two ocean crustal plates converge.

B When a plate of continental crust collides with a plate of oceanic crust, the heavier ocean crust gets subducted under the lighter and less dense continental crust. The process leads to fold mountains, fault mountains and volcanoes forming in the same area. Pressure from the subduction causes folding and faulting of sedimentary rocks. Some of the subducted oceanic crust melts and then rises and breaks through the crust of the earth causing volcanic eruptions and volcanic mountain formation on the surface. The Rocky Mountains, Andes Mountains and other mountains along the west coast of North and South America were formed in this way (see Case Study 2 on pages 244–245).

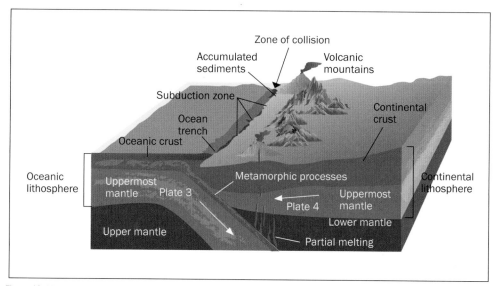

Figure 13 Mountains form where ocean crust and continental crust plates converge.

C When two plates of continental crust collide with each other, neither of them will subduct beneath the other due to their similar densities. Rock folding and faulting takes place in the areas of collision as the crustal rocks are crushed by the forces that cause the collision. Rock strata (layers of sedimentary rock) is bent and buckled into folds because of the great forces. Breaking (faulting) of the rock strata can also occur. These folding and faulting processes result in some rock being lifted higher than other rock. Many of the world's highest and largest chains of mountains were formed in this way including the European Alps and Himalaya Mountains in Asia (see Case Study 3 on pages 246–247).

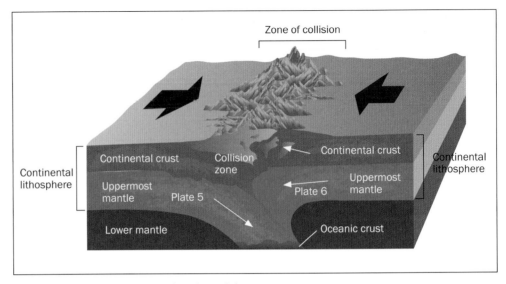

Figure 14 Mountains form where two continental crustal plates converge.

Some of the convergence took place hundreds of millions of years ago and the plates involved have joined together and the collisions zones are no longer active. Mountains formed this way are now not close to any active plate boundaries but are located in central areas of modern plates. The Appalachian Mountains in North America, the Scandinavian Mountains and Urals in Europe and the Great Dividing Range in Australia are mountains of this type. These mountains are so old that weathering and erosion has reduced them in height since they were first formed.

Learning Activities

1. Make a copy of Figure 11 and add information in each box A, B, C and D about events and processes that result in the formation of mountains. Box A has been completed as an example.

2. Why is there a close relationship between plate boundaries and the location of mountain areas?

3. Copy one of the three converging colliding boundary diagrams (A, B or C). Add to this diagram some more detail from the text and examples of mountain areas formed at this type of boundary.

4. What do these mountain ranges have in common — the Appalachian Mountains in North America, the Scandinavian Mountains and Urals in Europe, and the Great Dividing Range in Australia?

5. Mountain areas have an uneven global distribution. Write a paragraph explaining why. Include the names of mountains and mountain ranges from different parts of the world in your answer.

6. Describe what a 'linear pattern of mountains' means. Explain why mountain ranges often have a linear pattern (Figure 10 on page 239 has information that will help you answer this question).

 ISBN: 9780170233316

Case Study 1

Hot-spot volcanic mountains and the islands of Hawaii

Important understanding: Volcanic activity can lead to the formation of mountains.

The islands of Hawaii are hot-spot volcanic islands.

Hot-spots are places in the interior of the earth where there is a source of heat that causes bubbles and columns of magma to rise up and force their way through the crust, resulting in volcanic eruptions on the surface of the earth. These eruptions often take

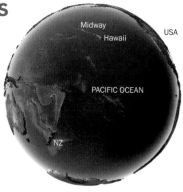

Figure 15 Hawaii and Midway Islands in the central Pacific.

place in central areas of plates far away from the boundary zones. This process leads to the formation of volcanic highland and volcanic mountains. The Hoggar and Tibesti Mountains in the Sahara Desert were formed this way (see page 236); so too were the islands of Hawaii.

If there is plate movement above the hot-spot, then a chain of mountains forms, and these get older as the distance from the hot-spot increases. Volcanoes formed this way have been described as riding on a conveyor belt. The Hawaiian islands in the central Pacific are at the southeastern end of a line of over 100 hot-spot volcanoes that began to form about 70 million years ago. The Hawaiian hot-spot has broken through the Pacific Plate with a line of volcanoes that are younger and higher towards the southeast (nearest to the hot-spot). As the islands formed, they got carried away from the hot-spot by the Pacific Plate moving northwest at a speed of 9 centimetres per year. Beyond Midway Island (Figure 15), the oldest of these volcanoes (in the Emperor group) have long ago subsided and been eroded beneath sea level. In Hawaii itself (Figure 16), Kauai, the island furthest from the hot-spot, formed around 5 million years ago, Oahu formed 3 million years ago, while on the 'Big Island' of Hawaii, the youngest of the Hawaiian islands, active hot-spot activity and mountain building continues (Figures 17–19). Beneath the ocean, about 35 km southeast of the Big Island, a new volcano called Lo'ihi is forming. At present it is 1000 metres below the ocean surface. It could form a new island above sea level within the next 100,000 years (Figure 16).

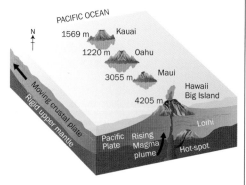

Figure 16 Volcanic hot-spot mountains — the islands of Hawaii.

Figure 17 Five-million-year-old eroded remains of the volcanic mountains of Kauai.

Figure 18 Volcanic mountain landforms of Maui, 2 million years old.

Figure 19 Kilauea hot-spot on the Hawaii 'Big Island' in eruption — present-day active mountain building.

Formation of the Andes – a collision area of oceanic and continental crust

Important understanding: The Andes mountains were formed because of plate collision and volcanic activity.

Formation of the present-day Andes started around 75 million years ago. Faulting, folding and uplifting of sedimentary and metamorphic rocks in the convergence area took place as the oceanic crust of the Nazca Plate was pushed against the stable rocks of the South American Plate (Brazilian Shield) to the east. The convergence forced the land up into high mountain chains and high plateau areas. In South America these high plateau areas are called 'altiplano'. Rising magma caused by melting of the subducting Nazca Plate resulted in the formation of volcanic chains within the mountains. With the plate movements and convergence still continuing, earthquakes and volcanic eruptions remain a feature of the Andes today. Since 1900 many magnitude 8 or greater earthquakes have occurred along the Nazca and South American Plate boundary, including a magnitude 9.5 earthquake in southern Chile in 1960, one of the largest ever recorded in the world.

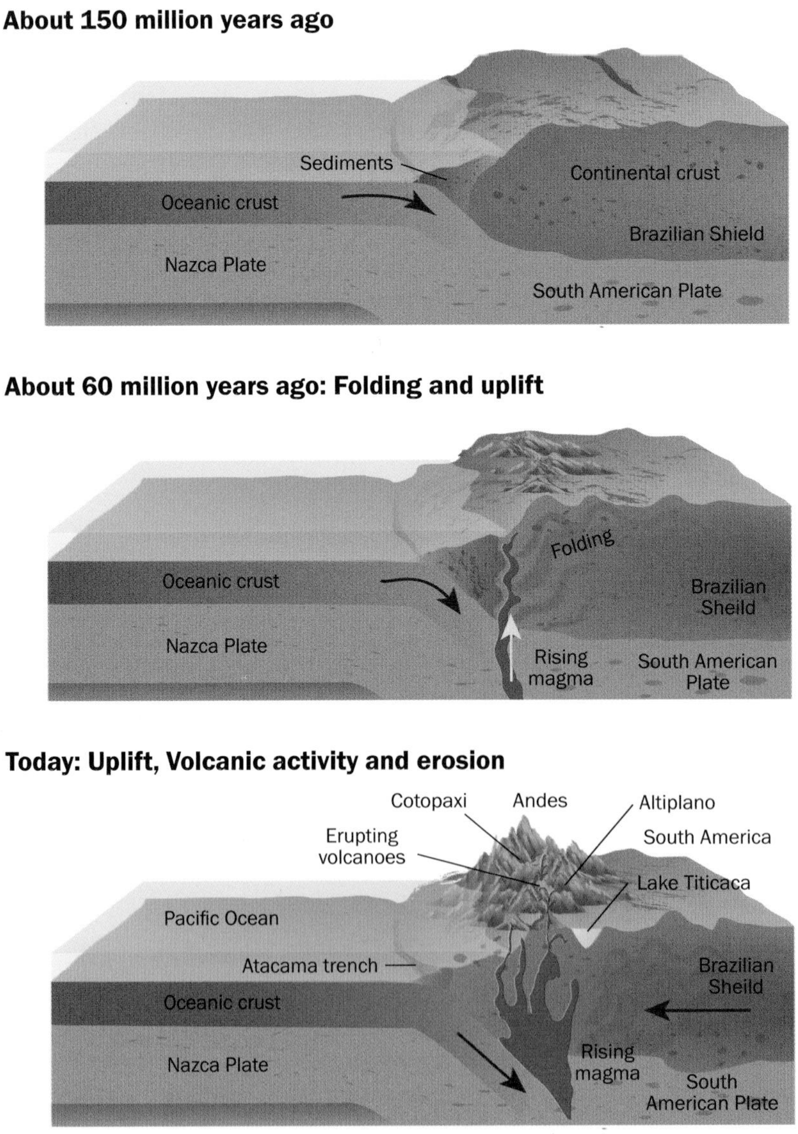

About 150 million years ago

About 60 million years ago: Folding and uplift

Today: Uplift, Volcanic activity and erosion

Figure 20 Colliding plates and formation of the Andes.

 ISBN: 9780170233316

Case Study 2

Towering Andes range formed quickly

Recent scientific research suggests the Andes Mountains formed more quickly than previously thought.

The Andes Mountains, the second-largest mountain belt in the world, formed between 10 and 6 million years ago — at a much faster rate than previously thought, according to recent research.

Traditionally, mountain formation has been described as a gradual folding and faulting of the earth's upper crust. Now, researchers are realising that the formation of gigantic mountains may occur far more quickly than once believed.

Carmala Garzione of the University of Rochester (USA) and colleagues reviewed studies of movements of the crust of the earth. Their findings suggest the formation of the Andes Mountains took place quickly when dense material beneath the earth's crust that formed the base of the mountains moved downward into the earth's mantle. This lightened the mountain base and caused the surface layers to rise up like a released cork. Garzione concluded that surface rocks rapidly rose a distance of between 1 and 3.5 kilometres over a period of just 1 to 4 million years.

Figure 21 Carmala Garzione working in the Andes.

Figure 22 Cotopaxi — andesitic volcanic cone (5897 metres) within the Andes Mountains in Equador (on the equator but high enough in altitude to have snow cover).

Case Study 3

Formation of the Himalayas – a collision of two continental land masses

Important understanding: Many of the rocks found high in the Himalayas were formed beneath the sea, and colliding continents have pushed these rocks upwards.

The world's greatest mountain range stretches 3000 km along the border between the Tibetan Plateau and India. The mountains are evidence of what happens when two continental land masses collide.

On 29 May 1953, New Zealander Edmund Hillary and Tenzing Norgay from Nepal completed the first successful climb of Mount Everest (8848 metres), the highest mountain in the Himalayas and the highest in the world. In parts of the Himalayas geologists have found the presence of sandstone and limestone rocks including shell and plant fossils that could only have formed beneath the sea. The sea has long since disappeared.

Two hundred million years ago a large land mass (now known as the Indian subcontinent) broke off from Africa and was carried northwards by plate movement towards the Eurasian continental plate. The sea called the Tethys Sea that existed between the two land masses became smaller and shallower as the two continents approached each other. The huge Himalayan mountain range began to form between 40 and 50 million years ago when the two land masses collided. The Tethys Sea drained away altogether as the land rose.

Because the two continental landmasses had about the same rock density, one plate could not be subducted under the other. Instead, the pressure of the two colliding plates could only be relieved by rocks being folded, faulted and pushed upwards in the collision zone. Rocks and marine fossils that had formed beneath the sea were lifted up and became part of the high mountains of the Himalayas. In addition to these sedimentary rocks, the great forces and heat associated with the collisions also led to the formation of metamorphic and igneous rock in the Himalayas. In mountain-building terms, the Himalayas have risen rapidly to a height of 8000–9000 metres in just 50 million years.

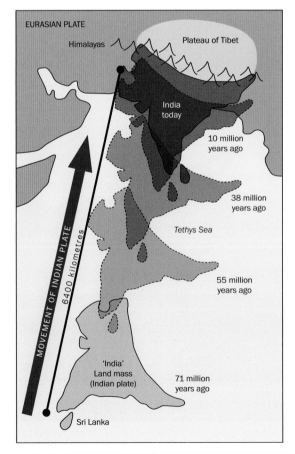

Figure 23 The Himalayas — 50 million years in formation.

Case Study 3

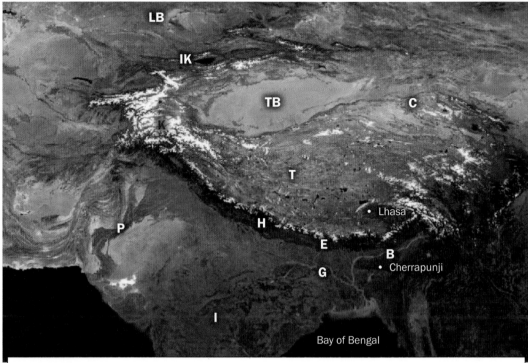

Place name key

LB	Lake Balkhash	TB	Tarim Basin	C	China	B	Brahmaputra River
K	Karakoram Mountains	T	Tibet	E	Mount Everest	G	Ganges River
P	Pakistan	H	Himalaya Mountains	I	India	IK	Lake Issyk Kul

Figure 24 The Himalaya area from space.

Figure 25 (a and b) Two views of Mount Everest in the Himalayas — ground level and satellite.

1 Case study 1: Hot-spot volcanic mountains and the islands of Hawaii

a Make a copy of both the map and diagram in Figure 16.

b Explain how the islands of Hawaii were formed.

2 Case study 2: Formation of the Andes — a collision area of oceanic and continental crust

a Draw an annotated sketch of Cotopaxi (Figure 22).

b What was the new idea that Carmala Garzione and her colleagues presented about how the Andes Mountains were formed?

3 Case study 3: Formation of the Himalayas — a collision of two continental land masses

a Make a copy of the satellite image in Figure 24. Add place names to the map using the information from the map key.

b Describe the location of the Himalaya Mountains.

4 Choose *one* of the three mountain case studies 1, 2 or 3. Write an illustrated report about this type of mountain (include at least one map, sketch or diagram in your report). Focus on explaining the location of the mountain type and how they were formed. Include the names of actual mountain areas in the report.

Mountain ranges and people – importance of mountain areas for people

Important understanding: Mountains and mountain areas are significant for the lives of people in many ways. They affect the lives of people directly, e.g. as tourist attractions, and indirectly, e.g. by influencing climate which in turn affects people.

- For many people mountains are their home. Millions of people across the world live in rural or urban mountain communities – on farms and in mountain villages, towns and cities.
- Interconnections and interdependence – mountains influence global rainfall, wind and temperature patterns. Lowland people depend on mountain environments for many products and services, including energy, timber, maintaining biodiversity and opportunities for recreation. Mountains are the source of the freshwater used by more than half of the world's population. Rivers that flow to the sea through lowland areas usually begin their life in mountain areas, so mountains, lowlands and sea are connected. The connection between mountain areas and areas beyond the mountains is two-way: many people visit mountain areas once, twice or on a regular basis. These people may visit for a day or for an extended period of time. These visitors provide jobs for mountain people and support for the economy of mountain areas.

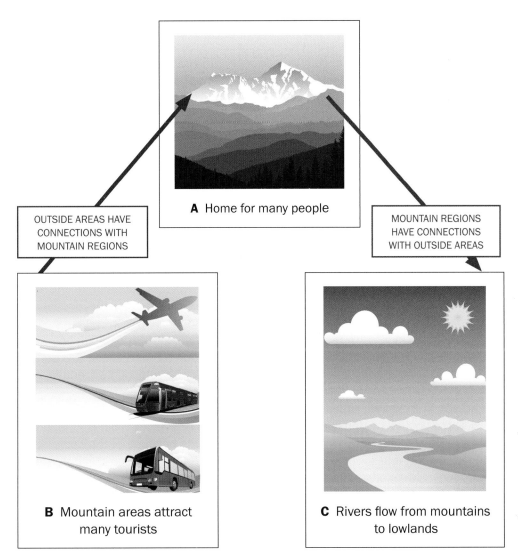

Figure 26 The significance and importance of mountains for people.

1 Make a large full-page or double-page copy of Figure 26. As you read through and study pages 249–268, add extra information and examples to each of the three boxes A, B and C about the significance and importance of mountain areas for people.

Home for millions people

Mountains cover 27 percent of the land surface of the world. Many mountain areas are places with very low population densities and little permanent settlement. Statistics, however, show that 12 percent of the world's population live in mountain areas – this is over 800 million people. If mountain areas were a country, it would be the third largest country in the world in terms of population total.

Seventy percent of all people living in mountain areas live in rural areas; the remaining 30 percent live in urban areas. Eighty-eight percent of the world's mountain population live in less-developed countries like Ethiopia and Peru, and in countries undergoing rapid transformation from being less developed to being developed like India and

China. The remaining 12 percent of the world's mountain population live in developed countries like Canada and Switzerland.

Population distribution within mountain areas is very uneven.
- 70 percent of the global mountain population live at lower altitudes at heights between 1000 and 1500 metres, 20 percent live between 1500 and 2500 metres and only 10 percent in the high mountains above 2500 metres. There are some huge areas that have very little or no settlement in them. These sparsely populated places are mostly the higher altitude areas above 3000 metres.

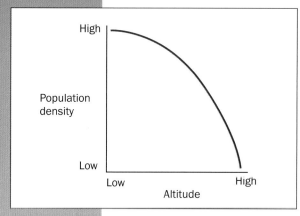

Figure 27 Population and altitude.

- In countries with a mix of mountains and lowlands, population distribution is very uneven. Population density decreases with height above sea level in these countries. People are concentrated in the lower areas. In Switzerland, for example, the Alps mountain region covers 60 percent of the land but contains only 11 percent of the population; the lower Jura Mountains cover 10 percent of the land and have 23 percent of the population living in them; while the lowest land of all, the Central Plateau covering 30 percent of Switzerland, has 66 percent of the population.

- However, in mountain areas of the tropics in Latin America, the Caribbean and Africa, the population density – altitude relationship is reversed: highest population densities are found between 2500 and 3500 metres because of the cooler less-humid climate and lower incidence of diseases in these places compared with surrounding lower areas. Bolivia in South America is an example of a country where more people live at higher altitudes than at lower ones. Sixty percent of the land of Bolivia is lowland but only 30 percent of the population live there. The other 70 percent live on the high Altiplano (high plateau) and mountain valleys of the Andes, which cover the remaining 40 percent of the country. A mild and moist climate and fertile soil provide good conditions for farming in many highland areas. Large and important cities like La Paz and Sucre are located in the highland region.

- Some well-known and large cities are mountain cities. Many of them are capital cities. North and South America, the Middle East and Asia all have many large mountain cities with populations over 100,000. Only in Europe and Oceania are there no really large cities located in mountains.

Zermatt, Switzerland:

Zermatt, a tourist town in the Swiss Alps. Most of the Swiss population live in mountain valleys and lowland towns and cities.

 ISBN: 9780170233316

La Paz, Bolivia:

Interesting fact: International soccer teams visiting Bolivia often blame the high altitude and lack of oxygen in the thin air for their low energy and dizziness when playing at the Hernando Siles stadium in La Paz (shown in Figure 29). For a short time in 2007, Bolivia was banned from playing international matches at the stadium because it was thought to give them an unfair advantage over visiting teams.

Large and well-known cities at high elevation

City	Country	Height of city (metres)	Population
La Paz	Bolivia	3640	880,000
Lhasa	China/Tibet	3490	375,000
Quito	Ecuador	2850	2,240,000
Bogota	Colombia	2619	7,400,000
Mexico City	Mexico	2420	8,850,000
Addis Ababa	Ethiopia	2362	2,750,000
Kabul	Afghanistan	1791	3,100,000
Nairobi	Kenya	1661	3,140,000
Kathmandu	Nepal	1400	1,000,000
Calgary	Canada	1048	1,100,000

Figure 28

Figure 29 La Paz (Bolivia), city in the Andes Mountains, over 3500 metres high.

Figure 30 La Paz, Bolivia.

ISBN: 9780170233316

Denver, Colorado, USA:

Interesting fact: The city of Denver (Colorado) in the USA has the nickname of 'The Mile High City'. The city is in the Rocky Mountains and has a population of 620,000. It gets its nickname because it has a location that is exactly one mile (5280 feet or 1609 metres) in elevation above sea level.

Figure 31 The flag of Denver and logo of the Denver Broncos.

Figure 32 Sports Authority Field at Mile High Denver, home of the Denver Broncos American Football Team.

Mountain areas: a source of minerals

- Mountain regions are the source of minerals that are important for individual countries and to the global economy. In the mid-19th century, gold discoveries in the mountains of western Canada and the USA like the Rockies led to the opening up of these areas and growth of mountain towns. The Andes Mountains in South America have a huge amount of mineral wealth (Figures 33 and 34). Silver and gold originally attracted mining companies into the mountains. Today Chile and Peru are the first and third largest producers in the world of copper – most of this is mined in the Andes and then taken to the coast by rail for export to overseas customers in the USA, Europe and Asia.

ISBN: 9780170233316

Figure 33 The huge Toquepala opencast copper mine in the Andes Mountains of Peru is 3500 metres above sea level. The pit is 2.5 kilometres across at the surface and is now more than 700 metres deep.

Figure 34 The mines at La Oroya are 4000 metres above sea level in the Andes Mountains in Peru. They produce copper, lead and zinc. The rail line carries the smelted minerals down from the Andes to the coastal port of Callao for export.

Learning Activities

1   Five questions to copy and answer:

a   As a general rule, which part of mountain areas is the most sparsely populated (has fewest people)?

b   An exception to the rule exists in some tropical mountain areas. What is this exception and what accounts for it?

c   Where is the Hernando Siles Stadium and why do many sports people not like playing and competing in this stadium?

d   In which two continents are there no very large mountain cities?

e   Name three minerals mined in the Rocky and Andes Mountains.

2   Draw pie graphs showing the percentage of world land area covered by mountains and the percentage of the world's population that live in these mountain areas.

3   On a map of the world, locate using proportional symbols or proportional bars the cities listed in Figure 28. Name each of the cities.

4   Draw an annotated sketch of either Figure 29 (La Paz) or Figure 33 (Toquepala).

5   Create an interesting infographic about the city of Denver.

6   Draw a set of bars to show the size of population of these top four countries. Add a fifth bar to that shows the number of people from across the world who live in 'mountain areas' (800 million people).

Top four countries by population size

Country	Population (millions)
China	1,362
India	1,236
USA	317
Indonesia	238

Impact on climate, water, soil and vegetation

Mountain areas are usually wetter and colder than nearby lowland areas.

• Moist air forced to rise as it meets a mountain barrier results in very heavy rainfall in the mountain area. Rain produced this way is known as relief or orographic rain. Lowlands on the sea-facing side of mountains are also usually wet, while rainfall on the inland side of mountains is usually very low. This pattern of low rainfall inland of mountains is called a rain-shadow effect – moisture from the air has been lost in rainfall as it has risen and crossed the mountains, and this results in an area of very dry climate inland of the mountains (Figure 35)

• Temperatures reduce as height increases – this helps to explain why snow-covered mountains are found on the equator in mountainous tropical areas like the Andes in Ecuador (Figure 22).

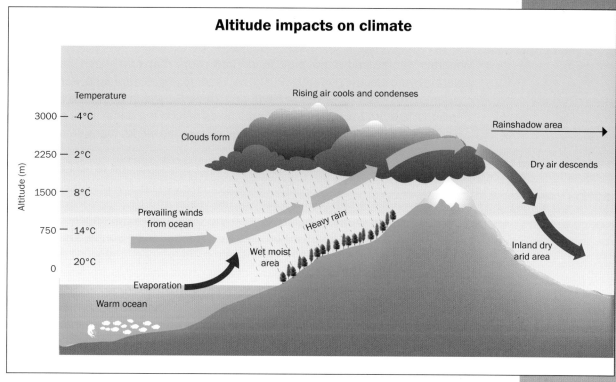

Altitude impacts on climate

Temperature

Rising air cools and condenses

Altitude (m)

3000 — -4°C

2250 — 2°C

1500 — 8°C

750 — 14°C

0 — 20°C

Clouds form

Rainshadow area

Dry air descends

Prevailing winds from ocean

Heavy rain

Inland dry arid area

Wet moist area

Evaporation

Warm ocean

Figure 35 Low temperatures, orographic rainfall and the rainshadow effect.

Case Study 4

World record – the wettest place on earth

The area around the town of Cherrapunji (also called Sohra) in the Khasi Hills region of northeast India lies around 1500 metres above sea level. This is the wettest area on earth – on average, 11,000 mm (11 metres) of rain falls here each year. In some years over 15,000 mm of rain has been recorded. Warm moist air from the Indian Ocean and Bay of Bengal flows into this area each summer (the summer monsoon). The air flows across the lowlands of Bangladesh and then hits the Khasi Hills where it is suddenly forced to rise up and over the steep slopes. The moisture-filled air forms clouds as the air condenses as it rises. Massive amounts of rain falls as the air moves north over the hills and on across the Himalayas. This is the orographic rainfall effect in operation. Water run-off and sediment eroded from the hills are carried down by rivers towards the sea through Bangladesh. Here the water is used for irrigation and helps build up a fertile soil when the rivers overflow their banks and deposit alluvium across the lowlands. Much further inland, the Plateau of Tibet and its major city Lhasa lie in the rainshadow of the Himalayas. This is a desert and semi-desert area with Lhasa receiving less than 500 mm of rain each year.

THE WETTEST PLACE ON PLANET EARTH CHERRAPUNJEE

A LAND OF BREATHTAKING BEAUTY & EXOTIC PEOPLE

CHERRAPUNJEE IS LOCALLY KNOWN AS 'SOHRA'.

THIS SIGN COURTESY

CHERRAPUNJEE HOLIDAY RESORT

PHONES: 03637-264218, 264219, 264220

E-MAIL: CHERRAPUNJEE@HOTMAIL.COM, CHERRAPUNJEE@YAHOO.CO.IN

WEBSITE: WWW.CHERRAPUNJEE.COM

AVERAGE ANNUAL RAINFALL (1973-2002) 12063.3MM

JAN, FEB, NOV & DEC, RECEIVE NOMINAL OR NO RAINFALL

Figure 36

High and regular rainfall plus snow melt in mountain areas provide both benefits and problems for people.

- Mountain rainfall and snow melt feeds into rivers that have been dammed to generate hydro-electric power (HEP) and for water to irrigate farmland. For example:
  - Giant dams like the Aswan in Egypt and Three Gorges in China are in lowland river valleys but the water that fills the lakes behind the dams flows down from the mountains. In mountain valley areas all over the world, many smaller hydro dams have been built (Figure 37).
  - In the Central Valley area of California in the USA, dams and canals have been built to take water from the Sierra Nevada and Coastal Range mountains into the dry southern San Joaquin part of the valley. The Central Valley, with less than 1 percent of the total farmland in the United States, produces 8 percent of the nation's agricultural output by value. It is the main source for a number of US food crops including tomatoes, almonds, grapes, cotton, apricots, and asparagus. This agricultural relies on water that originates from surrounding mountain areas (Figure 38).
- The high rainfall and steep slopes, however, are also a source of problems. Flooding, landslides and avalanches cause destruction, injury and death in a regular way to mountain communities. Sometimes the flood water and sediment pour out of the mountains and swamp farmland and communities in lowlands that lie below the mountains (Figure 39).

Figure 37 Cap-de-Long – small HEP dam and lake in the High Pyrenees mountains in southwest France.

Figure 38 Central Valley, California, an area of irrigated farmland using mountain water.

Figure 39 (a and b) Floods, avalanches and landslides in the Himalayas cause death, injury and disruption of lives.

Biodiversity

Height, climate and vegetation change over small distances in mountain areas. Travelling from the base to the top of a high mountain range like the Himalayas or Andes provides as much change in climate and vegetation as travelling from the equator to the North or South Poles. The term 'altitudinal zonation' is used to describe these changes – this is a concept that refers to the way different ecosystems occur at different altitudes. As altitude increases, there are changes in temperature, humidity, soil composition and solar radiation, and as a result distinctive vegetation and animal species also occur at different altitudes. In the mountain areas of Central and South America, four distinctive mountain zones are identified and known by their Spanish names.

Height (metres)	Name	Climate	Natural vegetation	Crops
0–1000	Tierra Caliente – warm and hot zone; zone of tropical crops	Hot (24°C–27°C) and humid	Luxuriant with dense forest in wetter areas	Bananas, sugar, cocoa
1000–1800	Tierra Templada – zone of temperate climate; zone of coffee crops	18°C–24°C	Open forest and grassland	Maize, coffee, tobacco
1800–3000	Tierra Fria – cold zone; zone of grain crops	12°C–18°C	Coniferous forest, grassland and scrub.	Wheat, barley, potatoes and other vegetables
Above 3000	Tierra Helada – zone of frost and frozen land	Under 12°C, often below freezing	Alpine plants and pasture with much bare rock, snow and ice	Llamas and alpacas on mountain pastures; few crops grown

Figure 40 Andes Mountains: altitundinal zonation.

The genetic diversity of mountains areas is an important resource in a changing world with a growing population. Mountain areas and the zones within them are the origin places of many plant and animal species. These include crops like maize, potatoes, barley, sorghum, tomatoes, and apples and animals like sheep, goats, yak, llama and alpaca. The original varieties of many of these crops and animals still found in the mountain areas are important in the breeding of new varieties and thereby help maintain food production. Mountain forests are also a source of timber for fuel and construction for mountain and lowland people. The Sierra Nevada mountain range in California in the USA provides an example of mountain area biodiversity. It has 10,000–15,000 different plants and animal species many of them unique to this area. Just north of here in the Cascade Range mountains of Washington state, where an important discovery was made in the 1960s. Scientists in a search for a cancer cure were collecting natural substances from plants. Thousands of substances were collected. One of these was from the bark of a tree called *Taxus brevifolia* (more commonly known as the Pacific yew tree). The substance became known as taxol and was found to be effective in the treatment of breast, ovarian and lung cancers. It has come to be widely used by doctors since then.

Mountain – lowland connections

Millions of people live on lowlands below mountain areas. Some can look up every day and see the mountains rising above their homes; for others the mountains are hundreds of kilometres away over the horizon. Mountains have an impact on the lives of all these people – mountains and lowlands are connected and interdependent in many ways. About 80 percent of the world's freshwater originates in the mountains. Heavy rainfall is a feature of mountain areas. Much of this runs off quickly into streams and rivers and flows to the lowlands. In some places the water remains locked up in mountains in snow, ice and lakes and gets released to the lowlands only slowly. All the world's largest rivers have their source in mountains.

Mountain water and soil is important for many lowland areas

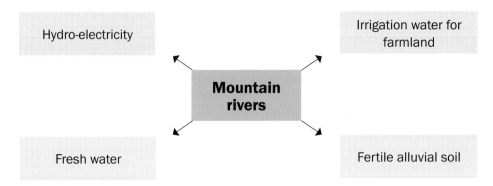

Figure 41

The high rainfall and in many places freezing temperatures as well cause rapid weathering and land erosion in mountain areas. Rivers carry the eroded material downstream into lowland areas. When rivers overflow their banks and flood the lowlands after heavy rain and snow melt in their mountain source areas, eroded material being carried downstream in the rivers is deposited across the lowlands. Over time, deep and fertile soil (called alluvium) gets built up – the mountains have come to the lowlands! Fertile soils that originated in the Himalayas cover the Ganges Valley in India and the Brahmaputra River lowlands of Bangladesh. River water that comes from the mountains is used each year to irrigate the lowland farms. Two great Chinese rivers, the Yangtze and Yellow, have their sources on the high plateau of Tibet. As a result of flooding and deposition over thousands of years by these two rivers, deep and fertile soils have built up across the lowlands of Eastern China. These fertile alluvial lowlands are home for millions of people. They are amongst the most densely populated and intensively farmed rural areas on the planet. The lives of two billion people in India, China, Pakistan and Bangladesh depend on water from rivers for drinking, for factory and farm use. This is mountain water that comes from the Himalayas and Plateau of Tibet.

Interesting fact: Cairo (Egypt), population 16 million, and Lima (Peru), population 8 million, are the world's two largest desert-located cities. Both cities rely on rivers that bring water from mountain source areas for their fresh water supply. The River Nile supplies Cairo, and the Rimac River brings water to Lima from the Andes.

Learning Activities

1   Make plus (positive) and minus (negative) lists showing the results and outcomes of the high rainfall that occurs in many mountain areas. Include place examples and mountain area names within your lists.

High amounts of rainfall in mountain areas	
+ Positive results and outcomes	**– Negative results and outcomes**
1 *Hydro-electric power generated in the Pyrenees Mountains*	1 *Death and injury from river floods in the Himalayas*
2	2
3	3

2   Describe the location of Cherrapunji. Explain what makes this small town and surrounding area so famous.

Learning Activities

3 Refer to the section 'Biodiversity'. Draw a diagram similar to this one that shows the idea of 'altitudinal zonation' in mountains like the Andes in South America.

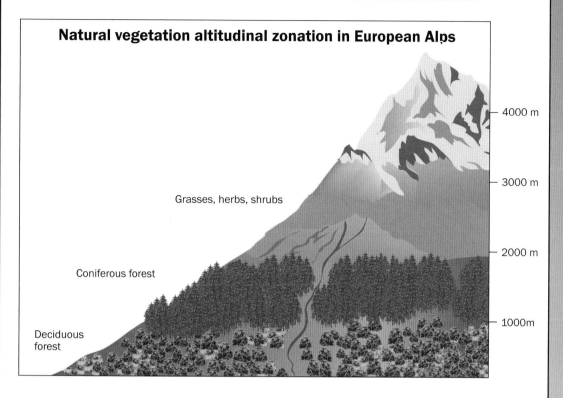

Natural vegetation altitudinal zonation in European Alps

Grasses, herbs, shrubs

Coniferous forest

Deciduous forest

— 4000 m

— 3000 m

— 2000 m

— 1000m

4 Why are plants and animals that are native to mountain areas so important for people? Include examples of named plants, animals and places in your answer.

5 Water from the River Nile supplies life-bringing water to millions of people and many places. This water has its origin in mountain areas. Use an atlas and/or online maps, images (e.g. Google Earth and Google Maps) plus other information to help you draw a map of the River Nile from source to mouth. Include on the map the following information.

- Where the River Nile begins. Hint: this will be in two mountain areas, one the source of the Blue Nile and the other the source of the White Nile.

- Where the River Nile flows as it travels from these source mountain areas to the sea.

- Include place, country and city names including the Tropic of Cancer, Sahara Desert, Egypt, Cairo, Khartoum, Aswan Dam and the Mediterranean Sea on your map.

- Title your map and give it a scale.

Mountains provide early warning of global climate change

Mountain regions are very sensitive to climate change. Scientists studying climate change get a lot of early warnings of global changes from making observations and measurements in mountain areas that help them understand what is happening to the global climate. Retreat of mountain glaciers and upward movement of the snowline have been observed from polar to tropical regions in recent decades. Rapid mountain glacier retreat has been recorded in Greenland, the European Alps, the Himalayas, Rocky Mountains, the Andes, New Zealand Southern Alps and in the mountain areas of East Africa (Figure 42).

Mountain glacier changes since 1970

Effective glacier thinning (m/yr)

| -1.4 | -1.2 | -1 | -0.8 | -0.6 | -0.4 | -0.2 | 0 | 0.2 | 0.4 | 0.6 | 0.8 | 1 | 1.2 | 1.4 |

Figure 42 Mountain glaciers are thinning and retreating across the globe (yellow and brown indicates thinning of the glacier ice, blue indicates thickening).

Glacier National Park is located in the Rocky Mountains in the state of Montana in the United States. The number of glaciers in the park has decreased from 125 in 1850 to less than 30 today. These remaining glaciers are estimated to be only a third as large as they were in 1850 (Figure 43). One study estimates that because of global warming, many of these remaining glaciers will disappear in the next 30 years.

Figure 43 Glacier retreat – Glacier National Park in the Rocky Mountains.

 ISBN: 9780170233316

Mountains act as barriers and boundaries

Mountains form a barrier to travel. In the past, mountain ranges like the Appalachians and Rockies in the United States and Great Dividing Range in Australia hindered the westward movement of new settlers and slowed the development of inland communities. The mountains also gave the North American Indians and Australian Aborigines some protection from the new migrants who were looking to settle land that was the traditional home of the indigenous peoples. Today, modern rail and rail routes have increased the accessibility into and through mountain areas, and air travel overcomes mountains as barriers to travel altogether.

The 'barrier' feature of mountains has led to them being used as geographical boundaries. One of the most famous boundary mountains are the Ural Mountains in Russia. These mountains have become seen as the feature that divides European Russia from Asian Russia. In other places, mountain ranges have been used as a natural feature that forms the national border. In South America, the high Andes form part of the boundary between Chile and Argentina, and in southern Europe, the Pyrenees Mountains form the boundary between France and Spain (Figure 44).

Figure 44 The Pyrenees Mountains, a national boundary.

Learning Activities

1 Refer to Figures 42 and 43.

    a What evidence is there that climate is changing?

    b What type of climate change does this evidence suggest is taking place?

2 Why would having early warning of climate change be of value for people?

3 Explain how mountains act as barriers and boundaries.

Tourism, sport and religion in mountain areas

Mountain tourism is a growing and important industry. After coasts and islands, mountains are the most important destination for tourists across the globe. About 15–20 percent of the global tourism industry is related to mountain areas. The European Alps account for 7–10 percent of this annual global tourism turnover, but modern forms of transportation, especially air travel, mean that parts of almost every mountain region across the world are tourist destinations. Tourists are attracted to mountains for many reasons, including the clean air and crisp climate, remoteness, spectacular landscapes and unique wildlife, local culture, history and heritage, and the opportunity to experience snow and participate in snow-based and other outdoor environment-related activities and sports. Activities range from sightseeing, photography, walking and tramping, skiing and snowboarding to a wide range of adventure tourism activities like canyoning and freefall paragliding.

Many mountain people and communities rely on tourism for jobs and income. Tourism growth as well as bringing change can have negative as well as positive consequences and outcomes for mountain people and mountain communities.

ISBN: 9780170233316

Negative consequences include:
- exposure to and adoption of foreign traditions, lifestyles and products that tourists bring with them, destroying the unique local culture, traditions and ways of life of mountain communities
- increased visitor numbers resulting in unsustainable use of scarce local resources such as firewood, fish and fresh water
- pollution of the environment from the waste, rubbish and noise generated by tourists
- jobs related to tourism being seasonal and providing little skill-building or training for local people
- profits from tourism being taken by companies based outside of the mountains, meaning little financial benefit to the local communities.

Well-managed tourism can have many positive outcomes:
- income from tourism can help reduce poverty and increase wealth in mountain communities
- to support tourism, infrastructure improvements like sealed roads, clean and piped water supply, sewerage systems, medical services, reliable electricity supply and Internet access, as well as benefiting the tourists, also improve the lives of local people
- employment and income from tourism can improve the self-sufficiency, survival and sustainability of mountain communities.

A quote from Nepali Sherpas in the 1980s: *'Tourists are like so many cattle, representing highly mobile, productive, and prestigious, but perishable, forms of wealth. Like cattle, tourists give good milk, but only if they are well fed.'*

The Mount Everest region of Nepal attracts over 20,000 trekkers (walkers and trampers) each year. Eighty percent of households in the Everest area get their income from tourism. However, the large number of trekkers means that 12 percent of the trail/track network is severely worn away and eroded. It is estimated that there is 17 tonnes of garbage generated by tourists along every kilometre of trail. About 25 percent of firewood consumption in the region is due solely to tourism – almost 1000 tonnes of firewood are burned daily during the peak tourist season in 225 campsites and tourist lodges. As forests in the national park around Everest are protected, high levels of deforestation take place outside the park. As well as a loss of the timber resource, this also leads to accelerated erosion of hillsides and rapid loss of precious topsoil.

Sport and challenge in mountain areas

Mountain areas are locations where many sports-based activities have been established and become popular. Some of these activities take place only in mountain areas, others take place outside of mountain areas but have developed in mountains in a special way. Many traditional farm-based rural mountain villages have been transformed into alpine tourist towns. Winter snow based sports attract many visitors to these towns.

Recent and future Winter Olympic Games locations and venues

Year	Host country	Host city	Alpine area for ski events
2006	Italy	Torino	Sestriere
2010	Canada	Vancouver	Whistler
2014	Russia	Sochi	Khuto Alpine Resort
2018	South Korea	Pyeongchang	Alpensia-Yongpyong

Figure 45

Mountain areas also attract 'sport tourists' at other seasons. Development and promotion of all-season tourism helps provide wealth and employment in mountain communities. Mountain biking and climbing are examples of such activities. High-altitude sports training centres have become a specialised industry in some mountain areas. For long-distance endurance events like the marathon, ironman and road cycling, the saying of 'train high, compete low' applies. Training at high altitude results in athletes developing a large concentration of red blood cells as the body adapts to the shortage of oxygen at high altitude. When returning to compete at low altitude, athletes have a competitive advantage. Many countries have set up training centres in mountains to help their athletes achieve peak performance in competition.

☐ swiss olympic | TRAINING BASE

At 1800 metres above sea level the Engadin is located at an ideal training altitude. Those who train here benefit from the dry, bracing alpine climate with an average of 322 days of sunshine every year.

The area is an official training centre for the Swiss Olympic Association and has a Swiss Olympic Sport Medical Base. Many top athletes and those training to become the next generation of top athletes use the natural conditions and optimum training facilities in the Engadin for their preparations.

Athletes are increasingly taking advantage of the performance-enhancing effects of training at high altitudes. Even just spending time at a height of between 1500 and 2200 metres above sea level has the same impact as a light training. Training sessions performed at these altitudes have been demonstrated to show greater effectiveness than training in the lowlands.

Mountain areas contain sites that are important religious places

These may be the peaks and summits themselves or shrines, temples and monasteries that have been built in the mountains.

The Himalayas, for example, are seen as the home of gods and centre of the world by many religions – the place that holds the underworld, earth and heavens together. The ancient Chinese saw the high mountain peaks as being props that held up the sky. Mount Kailash in Tibet is a sacred place to five religions: Buddhism, Jainism, Hinduism, Bon Po and Ayyavazhi. According to Hindu tradition, Kailash is the home of the god Shiva. Many people view Kailash as a paradise and the final destination of all souls, while others believe that Mount Kailash has supernatural powers that are able to clean the sins of a lifetime of any person. The great Buddhist monastery of Potala is in the high-altitude city of Lhasa in Tibet. Gangotri is a town located at a height of 3100 metres in the Indian Himalayas. Each year, 3000 mountaineers and porters and 30,000 trekkers visit the town. This number is dwarfed by the 300,000 Hindu pilgrims who visit. They come because the town is on the banks of the river Bhagirathi, which is the origin of the holy River Ganges and home of the goddess Ganga.

Figure 46 a–h: Mountain destinations in photos – tourism, sport and religion

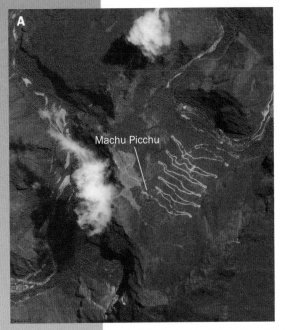

A and B Machu Picchu – Inca city in the Andes Mountains in Peru. A World Heritage site and Peru's most visited tourist attraction. The photos show the access path up from the valley, and the remains of the Inca buildings and terraces.

C Buddhist stupas (monuments) in Tibet, with holy Mount Kailash in the background.

D Swiss Alps mountain railway going from the tourist resort of Zermatt towards the Matterhorn (4478 metres).

E Tramping near Whistler (Canada) in the Coast Mountains.

F Skiing in the Carpathian Mountains of Ukraine.

G Mountaineering in the European Alps.

H Skiing in the Swiss Alps.

Learning Activities

1 Refer to pages 261–263 to help you answer these three questions.

a The mountains have always been there, so why have the number of tourists visiting mountain areas increased so much in the past 50 years?

b What would you judge to be the greatest positive and the greatest negative consequence of the great increase in the number of tourists for people who live in mountain areas? Justify your two answers.

c What do these four places have in common: Sestriere, Whistler, Khuto, Alpensia-Yongpyong?

d Design an advertisement to encourage athletes to book a training session at Engadin-St Moritz. Include the idea of 'train high, compete low' in the advertisement.

e Why are the Himalaya Mountains of religious importance to so many people?

2 Figure 46 has eight sets of photographs (A–H). Using information that is given both in the photo captions and in the text on pages 261–263, state the place/location of each photo and give reasons why people would travel to this location. For example: A – Machu Picchu in the Andes Mountains of Peru – to visit the old Inca city and see the remains of the buildings and terraces.

Mountains and people – the importance of high-elevation environments: Summary

Mountains are a common sight in every continent. Just as mountains are diverse, their importance is also diverse:

- They make up one-fifth of the world's landscape, and provide homes to at least one-tenth of the world's population.
- Two billion people depend on mountain ecosystems for most of their food, hydro-electricity, timber, and minerals.
- Mountains provide the freshwater needs of more than half of the world's population.
- Mountains provide opportunities for tourism, recreation, relaxation and spiritual renewal both for local people and those from far away.
- Mountains have ecological, socio-economic, and aesthetic value – not just for the people living on them, but also for the rest of the world as well.
- Mountain environments are essential for the survival of the global ecosystem.

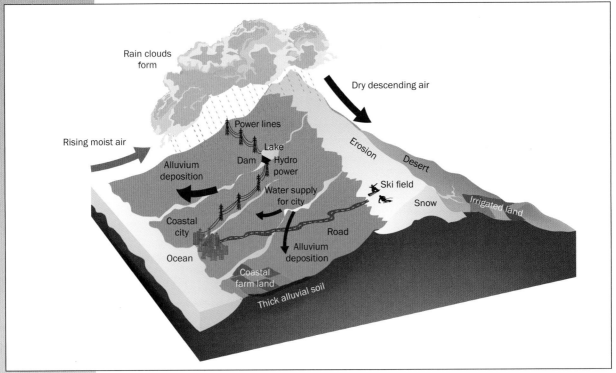

Figure 47 Mountain area – lowland area links.

Learning Activities

Choose one of these four activities.

EITHER

1. Use the bullet point list above to help you write a short essay titled 'The importance and significance of mountain areas for people'. Include examples of named places and named people within your answer.

OR

2. Make a copy of Figure 47. Number and name eight features on the diagram that show ways mountain areas impact on the lives of people. Make a key for the numbers beneath the map, stating the impact on the lives of people shown by each numbered feature, AND give an example of a place in the world where this happens. For example:

   a   Mountain area ski field, e.g. in the Carpathian Mountains of Ukraine

   b   Mountain water used to irrigate farmland, e.g. in the California Central Valley.

Learning Activities

OR **c** Based on the text and photo information on pages 248–266, create a crossword puzzle about 'The importance and significance of mountain areas for people'. Include a set of answers.

OR **d** Present a visual display that shows the importance of mountain areas for people.

'It was already busy when I went up Everest. Now it's like rush hour'

Adapted from an article by Steve Goodwin in the *Independent*, Saturday, 26 May 2012.

'As we climbed the steep ridge to the South Summit of Everest, the queuing was more like that of a supermarket check-out than a supposedly wild mountainside.' — this quote is from my report in the *Independent* of 23 May 1998.

Fourteen years on, and the queues have got worse. There were perhaps a dozen of us halted that bright mid-morning on the South Summit. At 8760 metres, we were less than 100 metres below Everest's main summit but owing to a mistake by another climbing team, nobody was going any higher that day.

This weekend, some 200 climbers hope to pass the point where we were forced to retreat, and go on to the top of the world. I wish them luck. Anyone who has endured the lung-searing grind for perhaps eight hours up from the camp on the South Col deserves their summit.

But 200 people? Even supposing the weather stays benign, that is a frightening number jostling on an icy aerial gangway, brains and bodies dulled by oxygen starvation and exhaustion. Plodding up, you're on autopilot, scarcely noticing the dawn light spreading across Tibet. The wind starts whipping the snow, you worry over the oxygen supply and you are dehydrated through not wanting to stop to drink and maybe lose your place in the ascending column. It's going to be a long way down. Perhaps it wasn't quite like this for those who have perished on Everest this season, but it is no surprise they died on the descent.

Is this really mountaineering? To Joe Public, since Everest is the world's highest mountain, it must be the ultimate goal of every true mountaineer. Far from it.

Given the chance of an Everest trip for free, many climbers would probably jump at it. But, handed NZ$60,000 (the typical cost of a 'climb Everest' package) and told to spend it climbing, the majority would rather go exploring, perhaps among the hundreds of 6000 metre-plus peaks in the Himalaya, Tibet and China that remain unclimbed.

Everest in the 21st century has become a mirror of the world that most of us go to the mountains to escape: a circus of people wanting an eye-rolling piece of dinner party one-upmanship, or a marketable line on a CV; media stunts to be the youngest, oldest, first blindfold on a tricycle, etc; and a world of instant electronic global communication.

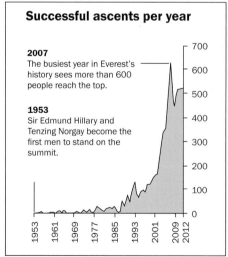

Successful ascents per year

2007
The busiest year in Everest's history sees more than 600 people reach the top.

1953
Sir Edmund Hillary and Tenzing Norgay become the first men to stand on the summit.

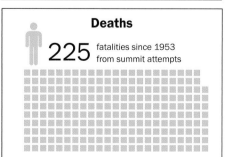

Deaths

225 fatalities since 1953 from summit attempts

Leanna Shuttleworth, a 19-year old woman who reached the summit this week, spoke of passing on her way up 'quite a few bodies . . . and even a couple who were still alive'; while the decision of one Israeli climber to abandon his attempt to rescue a fellow climber is considered so extraordinary as to make headlines. Meanwhile, three Sherpas died earlier in the season, preparing the route the client-climbers take through the Khumbu Icefall.

Perhaps while they are queuing at the Hillary Step, this weekend's Everest climbers might find time to examine their consciences.

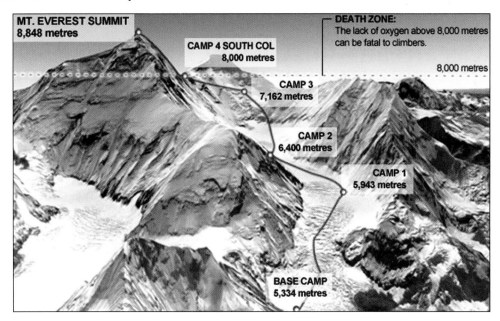

Learning Activities

1 Write a summary of this newspaper article in exactly 40 words.

2 What is the big concern being expressed in this article by Steve Goodwin? Include a copy of one of the three visuals to support your answer.

Index